De Gruyter Graduate

Klemeš, Varbanov, Wan Alwi, Manan • Process Integration and Intensification

Also of Interest

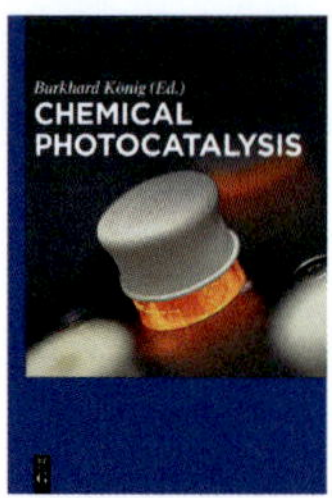

Chemical Photocatalysis, 2013
Burkhard König (Ed.)
ISBN 978-3-11-026916-1, e-ISBN 978-3-11-026924-6

Engineering Catalysis, 2013
Dmitry Yu. Murzin
ISBN 978-3-11-028336-5, e-ISBN 978-3-11-028337-2

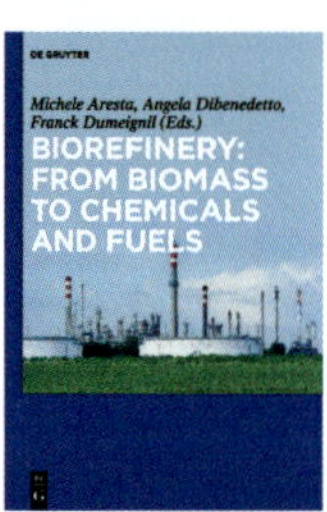

Biorefinery: From Biomass to Chemicals and Fuels, 2012
Michele Aresta, Angela Dibenedetto, Franck Dumeignil (Eds.)
ISBN 978-3-11-026023-6, e-ISBN 978-3-11-026028-1

Industrial Chemistry, 2013
Mark Anthony Benvenuto
ISBN 978-3-11-029589-4, e-ISBN 978-3-11-029590-0

Green Processing and Synthesis
Volker Hessel (Editor-in-Chief)
ISSN 2191-9542, e-ISSN 2191-9550

www.degruyter.com

Klemeš, Varbanov, Wan Alwi, Manan

Process Integration and Intensification

Saving Energy, Water and Resources

DE GRUYTER

Authors

Jiří Jaromír Klemeš
Centre for Process Integration and Intensifica-
tion – CPI², Research Institute of Chemical and
Process Engineering – MÜKKI
Faculty of Information Technology
University of Pannonia
Egyetem u.10, 8200 Veszprém
Hungary

Petar Sabev Varbanov
Centre for Process Integration and Intensifica-
tion – CPI², Research Institute of Chemical and
Process Engineering – MÜKKI
Faculty of Information Technology
University of Pannonia
Egyetem u.10, 8200 Veszprém
Hungary

Sharifah Rafidah Wan Alwi
Faculty of Chemical Engineering
Universiti Teknologi Malaysia
UTM Johor Bahru
81310 Johor
Malaysia

Zainuddin Abdul Manan
Faculty of Chemical Engineering
Universiti Teknologi Malaysia
UTM Johor Bahru
81310 Johor
Malaysia

ISBN 978-3-11-030664-4
e-ISBN978-3-11-030685-9

Library of Congress Cataloging-in-Publication data
A CIP catalog record for this book has been applied for at the Library of Congress.

Bibliographic information published by the Deutsche Nationalbibliothek
The Deutsche Nationalbibliothek lists this publication in the Deutsche Nationalbibliografie; detailed
bibliographic data are available in the Internet at http://dnb.dnb.de.

© 2014 Walter de Gruyter GmbH, Berlin/Boston
Typesetting: PTP-Berlin Protago-T$_E$X-Production GmbH, Berlin
Printing and binding: Hubert & Co., Göttingen
♾ Printed on acid-free paper
Printed in Germany

www.degruyter.com

Preface

Process Integration and Pinch Technology came to eminence in the late 1970's and early 1980's, and provided a phase change in the design of chemical processes, more especially in relation to energy use and energy efficiency. The central concept of this new approach was to provide targets for energy use in a chemical or related process, prior to the design required to achieve those targets. This new approach was not only elegant, but was rigorous in its thermodynamic basis, and provided a quick and systematic design of Heat Recovery in energy using processes.

The Centre for Process Integration in the University of Manchester Institute of Science and Technology (now part of The University of Manchester) became the focus of this new technology, which expanded to include efficient use of raw materials, emissions reduction, and operability. The research and applied technology developed at Manchester was transferred quickly to industry and academia worldwide by the use of professional development, graduate, and undergraduate programmes.

Professor Jiří Klemeš has been at the leading edge of this technology since its early pioneering days in Manchester. His efforts in promoting the Process Integration has been boosted by many researchers across the globe including Professor Petar Varbanov, Professor Sharifah Wan Alwi and Professor Zainuddin Abdul Manan who were themselves graduates of the Centre for Process Integration at UMIST.

Their contributions to the technology and its application are brought together in this exclusive book, which not only clearly examines and evaluates the many elements of the technology, but provides clear and concise examples of their application. The authors have provided all the elements required for complete understanding of the basic concepts in Heat Recovery and water minimisation in chemical and related processes, and followed these with carefully selected and developed problems and solutions in order to ensure that the concepts delivered can be applied. This book then extends this targeting to the design of Heat Recovery and water recovery networks, and again provides carefully selected problems and solutions.

Process Integration and Process Intensification are continuing to expand and evolve. The authors have included a final chapter which provides comprehensive and extensive insights into many of these new developments. Also included are sources of materials which can be easily accessed for those who wish to further enhance their knowledge of this important area of technology.

April 2014   Simon Perry
MSc Programme Director, Centre for Process Integration,
The University of Manchester, UK

Content

Acronyms, abbreviations and symbols

A	Heat transfer area, m^2
$A_{HEN,min}$	Minimum heat transfer area for a HEN, m^2
a_L	Regression parameter for the intercept ratio in a steam turbine, 1
A_{ST}	Intermediate regression parameter in the steam turbine model, MW
a_{tot}	Regression parameter for the total energy flow in a steam turbine, MW
b_0	Regression parameter in the steam turbine model, MW
b_1	Regression parameter in the steam turbine model, MW/°C
b_2	Regression parameter in the steam turbine model, 1
b_3	Regression parameter in the steam turbine model, 1/°C
BCCs	Balanced Composite Curves
BFW	Boiler feed water
b_L	Regression parameter for the intercept ratio in a steam turbine, 1/°C
B_{ST}	Intermediate regression parameter in the steam turbine model, 1
b_{tot}	Regression parameter for the total energy flow in a steam turbine, 1
C	Cooler / Cold utility
CC	Composite Curve
CCC	Cold Composite Curve
CHP	'Combined Heat and Power', usually meaning 'Combined Heat and Power Generation'
COND	Condenser
COP	Coefficient of performance for a thermodynamic machine – heat pump, refrigerator, etc., (-)
CP	Stream heat capacity flowrate, kW/°C or MW/°C
$C_{p,stream}$	Stream specific heat capacity, kJ/(kg · °C)
CP_C	Heat capacity flowrate of a cold stream, kW/°C or MW/°C
CP_H	Heat capacity flowrate of a hot stream, kW/°C or MW/°C
CW	Cooling Water
GCC	Grand Composite Curve
H [heat exchange]	Enthalpy flow, kW or MW
H [topology]	Heater / Hot utility
HCC	Hot Composite Curve
HEN	Heat Exchanger Network
HI	Heat Integration
$h_{IN,COLD}$	Inlet enthalpy of a cold stream, kJ/kg
$h_{IN,HOT}$	Inlet enthalpy of a hot stream, kJ/kg
$h_{OUT,COLD}$	Outlet enthalpy of a cold stream, kJ/kg
$h_{OUT,HOT}$	Outlet enthalpy of a hot stream, kJ/kg
HP (steam)	High pressure (steam)

h_s / h_{stream}	Heat transfer coefficient (film) of a process stream, $kW/(m^2 \cdot {}^{\circ}C)$
h_{STEXT}	Enthalpy of a steam turbine exhaust, MWh/t
h_{STIN}	Enthalpy of a steam turbine inlet, MWh/t
IP (steam)	Intermediate pressure (steam)
L	Steam turbine intercept ratio for calculation of the Willan's line coefficients, 1
LP (steam)	Low pressure (steam)
m_{COLD}	Cold stream mass flow rate, kg/s
MER	Maximum Energy Recovery
m_{HOT}	Hot stream mass flow rate, kg/s
m_{max}	Maximum steam mass flow through a steam turbine, t/h
MP (steam)	Medium pressure (steam)
m_{STEAM}	Steam flow, t/h
m_{stream}	Stream mass flow rate, kg/s
N_C	Number of cold streams
N_H	Number of hot streams
n_{ST}	Slope of the Willan's line for steam turbines, MWh/t
PI	Process Integration
PTA	Problem Table Algorithm
Q	Heat load/duty, kW or MW
Q_C	Utility cooling demand/load, kW or MW
Q_{Cmin}	Minimum utility cooling, in the context of a process-level Pinch Analysis, kW or MW
Q_{DIST}	Heat load associated with a distillation column, kW or MW
Q_H	Utility heating demand/load, kW or MW
Q_{HE}	Heat exchanger duty, kW or MW
Q_{Hmin}	Minimum utility heating, in the context of a process-level Pinch Analysis, kW or MW
Q_{REC}	Heat Recovery load, kW or MW
q_s / q_{stream}	Enthalpy change flow of a process stream or its part, kW or MW
Q_{sink}	Heat load supplied by a heat pump to a heat sink, kW or MW
Q_{source}	Heat load drawn by a heat pump from a heat source, kW or MW
Q_{STREAM}	Heat load of a process stream, kW or MW
$Q_{UC,above}$	Cold utility load above the Pinch, kW or MW
$Q_{UH,below}$	Hot utility load below the Pinch, kW or MW
REB	Reboiler
SUGCC	Site Utility Grand Composite Curve
T	Temperature, ${}^{\circ}C$
T*	Shifted temperature – mainly used in the process-level Pinch Analysis, for the Grand Composite Curve, ${}^{\circ}C$
T**	Double-shifted temperature – used for representing the plots for Total Site targeting, ${}^{\circ}C$

$T^\star_{PINCH}$	Pinch shifted temperature, °C
TB	Temperature boundary
$T_{COLD\ PINCH}$	Pinch temperature for the cold streams, °C
$T_{HOT\ PINCH}$	Pinch temperature for the hot streams, °C
TS	Total Site
T_S / T_{start}	Starting (supply) temperature, °C
$T_S^\star$	Shifted starting temperature, °C
T_{sink}	Temperature of a heat sink (of a heat pump), °C
T_{source}	Temperature of a heat source (of a heat pump), °C
TSP	Total Site Profile
T_T / T_{end}	Ending (target) temperature, °C
$T_T^\star$	Shifted target (end) temperature, °C
T_x	Unknown temperature to be calculated, °C
U	Overall heat transfer coefficient, $kW/(m^2 \cdot °C)$
U, W	Heat loads shifted via a HEN loop and HEN path, MW
VHP (steam)	Very High Pressure (steam)
W	Power, kW, MW
W_{INT}	Intercept of the Willan's line for steam turbines, MW
W_{MAX}	Maximum power from a steam turbine, MW
W_{total}	Total energy flow consumed from the fluid in a steam turbine, MW
XP	Cross-Pinch heat transfer, kW or MW
ΔH	Enthalpy difference (Enthalpy Flow), kW or MW
Δh_{is}	Isentropic enthalpy change across a steam turbine, MWh/t
ΔT	Temperature difference, °C
$\Delta T_{interval}$	Temperature difference of a temperature interval, °C
ΔT_{LM}	Logarithmic-mean temperature difference (for a heat exchanger), °C
ΔT_{min}	Minimum allowed temperature difference, °C
ΔT_{PUMP}	Temperature lift of a heat pump, °C
ΔT_{sat}	Saturation temperature drop across a steam turbine or an expansion zone, °C
η_{mech}	Efficiency of the mechanical transmission of a steam turbine (eventually including the power generator), 1

1 Process Integration and Intensification: an introduction

Numerous studies are being performed on improving the efficiencies of the supply and utilisation of energy, water and other resources whilst simultaneously reducing the pollutant emissions. This vital task is the focus of this textbook. It has been estimated that the majority of industrial plants throughout the world use up to 50 % more energy than necessary (Alfa Laval, 2011).

Usually reducing resource consumption is achieved by increasing the internal recycling and re-use of energy and material streams. Projects for improving process resource efficiencies can be very beneficial and also potentially improve the public perceptions of companies. Motivating, launching, and carrying out such projects, however, involves proper optimisation, based on adequate process models.

As a response to these industrial and societal requirements, considerable research efforts have been targeted towards Process Integration (PI) and Process Intensification (PIs). After a short assessment of these advanced engineering approaches, this handbook makes an attempt to describe the methodology that can lead a potential user through the introductory steps. This introduction provides a short overview of the historical development, achievements, and future challenges. After the introduction the text focuses on Process Integration as an important engineering tool that can be exploited to achieve the goals of Process Intensification.

1.1 Process Intensification

There have been several initiatives supporting the development in this area. One of them is the Process Intensification Network – PIN (PINET, 2013). This network declared that Process Intensification (PIs) was originally conceived in ICI (at that time, the Imperial Chemical Industries plc) as "the reduction of process plant volumes by two to three orders of magnitude". PIs targeted at that time the reduction of capital cost, primarily by minimising equipment installation factors, which involve piping, support structures etc. It has since become apparent that a rigorously pursued strategy of PIs has far wider benefits than mere CAPEX (Capital Expenditure, funds used by a company to acquire or upgrade physical assets such as property, industrial buildings or equipment) reduction, and its definition has accordingly been softened to include very significant plant size reductions based upon revolutionary or "step-out" new technology. PIs is not a mere evolutionary "apple-polishing" exercise of incremental development.

The benefits of PIs have extended far beyond the CAPEX reductions envisaged 35 years ago. The iron grip, which can now be imposed upon the fluid dynamic environment within a reactor, means that improved selectivities and conversions can usually be achieved.

However, the former ICI head of manufacturing technology Roger Benson said that after 30 years of false starts, the time has come for Process Intensification (PIs) technologies to make a major breakthrough in the chemical and process industries (IChemE, 2011).

The PIN web site summarises (PINET, 2013) the advantages as:

1. Better product quality
2. Just-in-time manufacture becomes feasible with ultra-short residence times.
3. Distributed (rather than centralised) manufacture may become economical
4. Lower waste levels reduce downstream purification cost and are conducive to "Green" manufacture
5. Smaller inventories lead to improved intrinsic safety
6. Better control of process irreversibilities can lead to lower energy use.

PIs can help companies and others meet all these demands, in the process industries and in other sectors. PIN network assists companies to compete, and helps researchers target successful research goals. The PIN and HEXAG (HEXAG and PIN, 2013) help students to gain an awareness of PIs for future use in their employment.

A kind of a bible of PIs has been published by Reay et al. (2008) and a very recent updated version (Reay et al., 2013). It covers main issues such as PIs as compact and micro-heat exchangers, reactors, intensification of separation processes, mixing, application areas in petrochemicals and fine chemicals, off-shore processing, miscellaneous process industries, the built environment, electronics and the home, specifying, manufacturing and operating PI plants.

Reay et al. (2013) correctly stated that the heat transfer engineer notes that 'intensification' is analogous to 'enhancement', and intensification is based to a substantial degree on active and, to a lesser extent, passive enhancement methods that are used widely in heat and mass transfer.

There has been also a Working Party on Process Intensification (2013). Their mission statement declares:

"Process Intensification presents one of the most significant developments in chemical and process engineering of the past decennia. It attracts more and more attention of the chemical engineering community. Several international conferences, smaller symposia/workshops every year, books and a number of dedicated issues of professional journals are a clear proof of it. Process Intensification with its ambition and ability to make chemical processing plants substantially smaller, simpler, more controllable, more selective and more energy-efficient, addresses the fundamental sustainability issues in process industry and presents the core element of Green Chemical Engineering. In many research centres throughout Europe and the world numerous PIs-oriented research programs are carried out. Process Intensification is taught at various courses and gradually enters the regular university curricula. In the UK and in the Netherlands national PIs-networks have been operating for a number of years. Similar network is being formed in Germany (DECHEMA, 2013). Process Intensification plays an important role in the CEFIC's Technology Platform on Sustainable Chemistry (CEFIC, 2013). Process Intensification has now established its organisational position within the European Federation of Chemical Engineering."

The PIs WP has been organising annual conferences called European Process Intensification Conference (EPIC, 2013). The previous EPIC conferences were held in Copenhagen – 2007, Venice – 2009 and Manchester – 2011.

1.2 Process Systems Engineering and Process Integration

PIs has been very much targeted at processing units. On the other hand, Process Integration (PI) has developed a methodology by which PIs can be very benefitial at the system level. PI has been an important part of Process Systems Engineering, which is handled by the Working Party of Computer Aided Process Engineering of the European Federation of Chemical Engineering (2013). It has been one of the longest serving working parties. Its definition reads as:

> "Computer Aided Process Engineering (CAPE) concerns the management of complexity in systems involving physical and chemical change. These systems usually involve many time and scale lengths and their characteristics and the influences on them are uncertain. CAPE involves the study of approaches to analysis, synthesis, and design of complex and uncertain process engineering systems and the development tools and techniques required for this. The tools enable process engineers to systematically develop products and processes across a wide range of domains involving chemical and physical change: from molecular, thermodynamic and genetic phenomena to manufacturing processes and to related business processes. The Working Party promotes the development, study, and use of CAPE tools and techniques.
>
> Main characteristics are: Multiple scales, uncertainty, multidisciplinary, complexity.
>
> Core competencies: Modelling, synthesis, design, control, optimisation, problem solving Domains: manufacturing products and processes involving molecular change, sustainability, business processes, biological systems, energy, water."

The CAPE Working Party main venue has been ESCAPE – European Symposium of Computer Aided Process Engineering, whose first venue was organised already in 1992 in Helsingør, Denmark. The 2014 ESCAPE conference is in Budapest (ESCAPE 24, 2014), Copenhagen in 2015 and Portoroz in Slovenia in 2016. The CAPE WP has also collaborated with EURECHA to organise the CAPE Forum.

PI has roots as deep as 1972 and has been pioneered by several research centres, originally in the UK (UMIST Manchester), Japan and the US. However, over the years the methodology and research have spread out across the world. During the last years especially Asian researchers have become very active and are more and more leading the effort. There have been numerous publications devoted to PI for the more than 40 years of development covering methodology, and industrial implementations.

Also a number of excellent reviews have been presented. The most recent is the Handbook of Process Integration (Klemeš, 2013). The vast majority of leading PI researchers contributed with their unique expertise to this handbook.

PI has had a dedicated conference for 17 years already: PRES – Process Integration, Modelling and Optimisation for Energy Saving and Pollution Reduction. The next conferences are in 2014 in Prague (PRES, 2014) and 2015 in Sarawak, Borneo, Malaysia.

1.3 Contributions to PIs and PI to energy and water saving

To save energy on a large scale and on a global basis, companies taking ownership and responsibility for new plants clearly need to question the energy efficiency of the equipment and layout recommended to them by proposing contractors. Historically, designers and builders of process plants have not been asked or paid to critically review the energy efficiency options available to their clients, preferring to offer low risk, easy-to-replicate and therefore easy-to-guarantee generic designs. For existing plants, reducing energy consumption can be more challenging than is the case for grassroots developments.

PI supporting Process Design, Integration, and Optimisation has been around for more than 40 years (Klemeš and Kravanja, 2013). Its on-going development has been closely related to the development of Chemical, Power and Environmental Engineering, the implementation of mathematical modelling, and the application of information technology. Its development has accelerated over the years as its methodology has been able to provide answers and support on important issues regarding economic development – better utilisation and savings regarding energy, water, and other resources.

1.4 What is Process Integration?

Process Integration (PI) is a family of methodologies for combining several parts of processes or whole processes to reduce consumption of resources or harmful emissions into the environment. It started mainly as Heat Integration (HI) stimulated by the energy crises of the 1970s. HI has been extensively used in the processing and power generating industries over the last 40 years. PI examines the potential for improving and optimising the Heat Exchange between heat sources and sinks aimed at reducing the amount of external heating and cooling utilities, together with the related costs and emissions. It provides a systematic design procedure for energy recovery networks.

HI (using Pinch Technology) has several definitions, almost invariably referring to the thermal combination of steady-state process streams or batch operations for achieving Heat Recovery via Heat Exchange. More broadly, the definition of PI, as adopted by the International Energy Agency (Gundersen, 2000) reads:

> *'Systematic and general methods for designing integrated production systems ranging from individual processes to Total Sites, with special emphasis on the efficient use of energy and reducing environmental effects.'*

Reducing an external heating utility is usually accompanied by an equivalent reduction in the cooling utility demand (Linnhoff and Flower, 1978). This also tends to reduce the CO_2 emissions from the corresponding sites. Reduction of wastewater effluents based on Water (Mass) Integration can also lead to reduced freshwater intake (Wang and Smith, 1994), as demonstrated in industrial implementations elsewhere (Thevendiraraj et al., 2003).

1.5 A brief history of the development of Process Integration

Several methodologies emerged in late 1970s as a response to these industrial and societal challenges. One of them was "Process System Engineering" (PSE) (Sargent, 1979) and later extended again by Sargent (1983). Another methodology that received world prominence was "Process Integration" (PI) (Linnhoff and Flower, 1978). PI was first formulated within the first PI book presented by Linnhoff et al. (1982). Its development was further contributed to by a number of works from UMIST, Manchester, UK and other research groups.

It is remarkable that PI has never lost the interest of researchers during the last 40 years and has even been flourishing recently. HI has proved itself to have a considerable potential for reducing the overall energy demand and emissions across a site, leading to a more effective and efficient site utility system. One of the first related works was that by Hohmann (1971) in his PhD thesis at the University of Southern California. This work was the first to introduce systematic thermodynamics-based reasoning for evaluating the minimum energy requirements of a Heat Exchanger Network (HEN) synthesis problem. Various approaches dealing with the optimum HEN synthesis have since been published. Some of them became very popular, such as the work by Ponton and Donaldson (Ponton and Donaldson, 1974). The comprehensive overview of HEN synthesis presented by Gundersen and Naess (1990) and the overview of process synthesis presented earlier by Nishida et al. (1977) provided considerable impetus for the further research and development in this field – as can be witnessed in the more recent overview by Furman and Sahinidis (2002).

In some literature sources such as e.g. (Gundersen, 2000) it is declared that the concept of HI based on Recovery Pinch was independently discovered by Hohmann (1971) and by two research groups: (i) the two-part paper by Linnhoff and Flower (1978) and Flower and Linnhoff (1978), followed up by the Linnhoff PhD thesis (1979) and (ii) by the group around Umeda et al. (1978).

Gundersen (2000) also stated that Hohmann (1971) was the first to provide a systematic way of obtaining energy targets by using his Feasibility Table. Hohmann (1971) completed his PhD, within which some basic principles were included. However, with the exception of a conference presentation (Hohmann and Lockhart, 1976), he has never extensively published the results in a way that would attract a wider audience. Moreover, a lesser known important part leading to the Problem Table Algorithm was published at that time by MSc student Bodo Linnhoff at ETH – Zürich (Linnhoff, 1972). During those pre-information technology times, interactions amongst researchers were slower and more difficult. It was only possible to discover what other researcher were working on after printed publications were made available.

During the remaining part of the 1970s, Bodo Linnhoff, who was at that time a PhD student at the University in Leeds again perused and realised the potential of HI. The beginning was difficult, as his first paper (Linnhoff and Flower, 1978), which later became very highly cited, was nearly rejected by a leading journal of that time. Bodo's strong will and persistence prevailed in getting the idea published and off the ground. After the first paper's initial difficult birth, others followed – Flower and Linnhoff (1978) and Flower and Linnhoff (1979) followed smoothly.

The other group that produced interesting contributions was from Japan – at the Chiyoda Chemical Engineering & Construction Co., Ltd. Tsurumi, Yokohama. They published a series of publications on HEN synthesis – Umeda et al. (1978, 1979), optimum water re-allocation in a refinery – Takama et al. (1980), and applications of the T-Q diagram for heat-integrated system synthesis – Itoh at al. (1986).

The publication of the first "red" book by Linnhoff et al. (1982) played a key role in the dissemination of HI. More recently this book received a new extensive Foreword (Linnhoff et al., 1994) and content update. This User Guide through Pinch Analysis provided an insight into the more common process network design problems, including Heat Exchanger Network synthesis, Heat Recovery targeting, and selection of multiple utilities. As a spin-off from the Leeds centre, further works originated in Central Europe have been published. Firstly Klemeš and Ptáčník (1985) presented an attempt to computerise HI (Ptáčník and Klemeš, 1987), followed by HEN synthesis development and on mathematical methods for HENs (Ptáčník and Klemeš, 1988).

The full-scale development and application of these methodologies has been pioneered by the Department of Process Integration, UMIST (now the Centre for Process Integration, CEAS, at The University of Manchester) in the 1980s and 1990s. Amongst other earlier key publications was Linnhoff and Hindmarsh (1983) with presently more than 450 citations in SCOPUS, followed by a number of works dealing with extensions, e.g. the summary by Smith et al. (1995), first updated Russian version (Smith et al., 2000) and more updated version Smith (2005). A specific food industry overview of HI was presented by Klemeš and Perry (2008) in a book edited by Klemeš et al. (2008) and recently by Klemeš et al. (2010). Tan and Foo (2007) successfully applied the Pinch Analysis approach to carbon-constrained energy sector planning. More recently Foo et al. (2009) applied the Cascade Analysis technique to carbon and footprint-constrained energy planning, while Wan Alwi and Manan (2008) provided a holistic framework for designing cost-effective minimum water utilisation networks.

Setting targets for HI has been widely publicised by Linnhoff et al. (1982), followed by two books of Smith (1995) and the updated version Smith (2005). Smith at al. (2000) presented a Russian version extended by a number of examples and the reader, who can read in this language, is encouraged to look at this source. The second edition of Linnhoff et al. (1994) was elaborated by Kemp (2007).

A very good analysis has been developed by Gundersen (2013) in his chapter of the PI Handbook edited by Klemeš (2013). Gundersen (2013) summarised the important elements in basic Pinch Analysis as:
(a) Performance Targets ahead of design,
(b) The Composite Curves representation can be used whenever an "amount" (such as heat) has a "quality" (such as temperature), and
(c) The fundamental Pinch Decomposition into a heat deficit region and a heat surplus region.

Heat Integration has been extended into various directions as Total Site Heat Integration (Dhole and Linnhoff, 1993) further extended by Klemeš et al. (1997) and to Locally Integrated Energy Systems (Perry et al., 2008). Mohammad Rozali et al. (2012) extended the Heat Integration on Total Site by including the power (electricity) into the game as well.

Direct implementation of Process Intensification by heat transfer intensification pioneered by Zhu et al. (2000) was extended by Kapil et al. (2012). It opened another exciting avenue for PI. Pan et al. (2012) implemented this methodology to industrial plants retrofit.

However, very important for all design and optimisation problems are the quality of the input data and their verification. Manenti et al. (2011) contributed to data reconciliation, which is one of key issues to mine the faithful data representing the real plant. Klemeš and Varbanov (2010) provided one of surprisingly not too many published so far advices about potential pitfalls of Process Integration.

The most obvious analogy to heat transfer is provided by mass transfer. In heat transfer, heat is transferred with temperature difference as the driving force. Similarly, in mass transfer, mass (or certain components) is transferred using concentration difference as the driving force. The corresponding Mass Pinch concept, developed by El-Halwagi and Manousiouthakis (1989), has a number of industrial applications whenever process streams are exchanging mass in a number of mass transfer units, such as absorbers, extractors, etc.

Mass Integration (MI) is a branch of PI providing a systematic methodology for optimising the global flow of mass within a process based on setting performance targets and for optimising the allocation, separation, and generation of streams and species. Within the context of wastewater minimisation, a MI problem involves transferring mass from rich process streams to lean process streams in order to achieve their target outlet concentrations whilst simultaneously minimising the waste generation and the consumptions of the utilities – including freshwater and external mass separating agents. On this topic El-Halwagi has produced three valuable books on PI; the first focuses on pollution prevention through PI (El-Halwagi, 1997), the second is a more general discussion on PI and including MI (El-Halwagi, 2006) and the most recent book investigates sustainable design through PI (El-Halwagi, 2012).

The process network synthesis associated with these chemical properties cannot be addressed by conventional MI techniques, so another generic approach has been developed to deal with this problem firstly by Shelley and El-Halwagi (2000) and further extended by El-Halwagi et al. (2003). For systems that are characterised by one key property, Kazantzi and El-Halwagi (2005) introduced a Pinch-based graphical targeting technique that establishes rigorous targets for minimum usage of fresh materials, maximum recycling, and minimum waste discharge.

One specific application of the Mass Pinch is in the area of Wastewater Minimisation, where the optimal use of water and wastewater is achieved through reuse, regeneration and possibly recycling. The corresponding Water Pinch, developed by Wang and Smith (1994), can also be applied to the design of Distributed Effluent Treatment processes. A substantial work has been delivered by the UTM group as Wan Alwi and Manan (2006) developed SHARPS – A New Cost-Screening Technique To Attain Cost-Effective Minimum Water Utilisation Network, Wan Alwi and Manan (2010) STEP – A new graphical tool for simultaneous targeting and design of a Heat Exchanger Network and Wan Alwi at al. (2010) a new graphical approach for simultaneous mass and energy minimization.

An interesting extension has been Process Integration for Resource Conservation (Foo, 2012).

1.6 The aim and scope of this textbook

This handbook is more focussed on the detailed description of the selected principles to the extent that the reader should be able to solve both the illustration problems and also the real life industrial task.

It has been supplemented by a number of hands-on working sessions to allow readers to practise and deepen the understanding of the problem and to avoid potential pitfalls during the solution. The working sessions have been based on the authors' long-term academic teaching experience as well as on delivering further development courses for the industry worldwide.

The authors would like to express a deep gratitude to all colleagues at UMIST, The University of Manchester, Universiti Teknologi Malaysia, University of Pannonia, University of Nottingham Malaysia Campus, University of Maribor, Brno University of Technology and many others – fellow staff members, postgraduate and even undergraduate students who by using this material for their education provided most valuable feedbacks and comments, which substantially contributed to the testing, verification and the development of the handbook material. A special acknowledgement should be provided to the delegates from the industry, whose comments have been most valuable. The authors would like to mention at least the two of those most recent – MOL Hungarian Oil Company and PETRONAS Malaysia.

References

Alfa Laval (2011). Don't think of it as waste – it's energy in waiting, local.alfalaval.com/en-gb/aboutus/news/Pages/WasteHeatRecovery.aspx accessed 02/07/2013.

CAPE (2013). www.cape-wp.eu, accessed 12/07/2013.

CEFIC (2013). The European Chemical Industry Council, www.cefic.org, accessed 16/04/2013.

DECHEMA (2013). Gesellschaft für Chemische Technik und Biotechnologie e.V. (Society for Chemical Engineering and Biotechnology) www.dechema.de/en/start_en.html accessed 12/04/2013.

Dhole, V.R. and Linnhoff, B. (1993). Total site targets for fuel, co-generation, emissions and cooling, *Computers & Chemical Engineering*, 17(Suppl.), 101–109.

El-Halwagi M.M. (1997). *Pollution Prevention through Process Integration: Systematic Design Tools*, San Diego, USA: Academic Press.

El-Halwagi, M.M. (2012). Sustainable Design through Process Integration – Fundamentals and applications to industrial pollution prevention, resource conservation, and profitability enhancement, Elsevier. Amsterdam, The Netherlands. www.knovel.com/web/portal/browse/display?_EXT_KNOVEL_DISPLAY_bookid=5170&VerticalID=0 accessed 01/07/2013.

El-Halwagi, M.M. and Manousiouthakis, V. (1989). Synthesis of mass-exchange networks, *American Institute of Chemical Engineering Journal*, 35, 1233–1244. DOI: 10.1002/aic.690350802.

El-Halwagi, M.M. (2006). *Process Integration*. Amsterdam, Netherlands: Academic Press.

El-Halwagi, M.M., Gabriel, F. and Harell, D. (2003). Rigorous graphical targeting for resource conservation via material recycle/reuse networks, *Industrial and Engineering Chemistry Research*, 42 (19) 4319–4328.

EPIC (2013). European Process Intensification Conference, www.chemistryviews.org/ details/ event/2502531/EPIC_2013_–_European_Process Intensification_Conference _EPIC_ 2013.html accessed 06/07/2013.

ESCAPE 24 (2013). European Symposium on Computer Aided Engineer, Budapest, Hungary www.escape24.mke.org.hu/home.html accessed 17/04/2014.

Flower, J.R. and Linnhoff B. (1978). Synthesis of Heat Exchanger Networks – 2. Evolutionary generation of networks with various criteria of optimality, *AIChE J* 24(4) 642–654.

Flower, J.R. and Linnhoff, B. (1979). Thermodynamic analysis in the design of process networks, *Computers and Chemical Engineering*, 3 (1–4), 283–291.

Foo, D.C.Y. (2012). *Process Integration for Resource Conservation*, Boca Raton, Florida, USA: CRC Press.

Foo, D.C.Y. (2009) State-of-the-art review of pinch analysis techniques for Water network synthesis, *Industrial and Engineering Chemistry Research*, 48 (11), 5125–5159.

Furman, K.C. and Sahinidis N.V. (2002). A critical review and annotated bibliography for heat exchanger network synthesis in the 20th century, *Ind. Eng. Chem. Res.* 41(10), 2335–2370.

Gundersen, T. (2000). *A process integration primer – Implementing agreement on process integration*, Trondheim, Norway: International Energy Agency, SINTEF Energy Research.

Gundersen, T. (2013). Heat Integration: Targets and Heat Exchanger Network design, Chapter 4. In: Klemeš J.J. (ed.), *Handbook of Process Integration (PI): Minimisation of Energy and Water Use, Waste and Emissions*, Cambridge, United Kingdom: Woodhead Publishing, ISBN-13: 978 0 85709 593 0.

Gundersen, T. and Naess, L. (1990). The synthesis of cost optimal heat exchanger networks: An industrial review of the state of the art, *Heat Recovery Systems and CHP*, 10(4), 301–328.

HEXAG and PIN online newsletter, (2012), www.hexag.org/newsletter/index.html accessed 11/04/2013.

Hohmann, E. and Lockhart, F. (1976). Optimum heat exchangers network synthesis, in: *Proceedings of the American Institute of Chemical Engineers*, American Institute of Chemical Engineers, Atlantic City, NJ, USA.

Hohmann, E.C. (1971). *Optimum networks for heat exchange*, PhD thesis, University of Southern California, Los Angeles, USA.

IChemE, (2011). Advanced Chemical Engineering Worldwide, News archive, 21 June 2011, www. icheme.org/media_centre/news/2011/process%20intensification%20the%20time%20 is%20now.aspx#.UXBG-kpChuM accessed 24/06/2013.

Itoh, J., Shiroko, K, and Umeda, T, (1986) Extensive applications of the T-Q diagram to heat integrated system synthesis, *Computers & Chemical Engineering*, 10 (1) 59–66.

Kapil, A., Bulatov, I., Smith, R., and Kim, J.-K. (2012). Process integration of low grade heat in process industry with district heating networks, *Energy*, 44(1), 11–19.

Kazantzi, V. and El-Halwagi, M.M. (2005). Targeting material reuse via property integration, *Chemical Engineering Progress*, 101 (8), 28–37.

Kemp, I.C. (2007). *Pinch Analysis and Process Integration. A User Guide on Process Integration for Efficient Use of Energy*, (authors of the first edition Linnhoff, B., Townsend, D.W., Boland, D., Hewitt, G.F., Thomas, B.E.A., Guy, A.R. and Marsland, R.) Amsterdam, the Netherlands: Elsevier.

Klemeš, J. (ed.) (2013). *Handbook of Process Integration (PI)*, Woodhead Publishing Series in Energy No. 61, Cambridge, UK & Philadelphia, USA: Woodhead Publishing Limited.

Klemeš, J.J. and Kravanja, Z. (2013). Forty years of Heat Integration: Pinch Analysis (PA) and Mathematical Programming (MP), Current Opinion in Chemical Engineering, DOI: 10.1016/j.coche.2013.10.003.

Klemeš, J., Dhole, V.R., Raissi, K., Perry, S.J., and Puigjaner, L., (1997).Targeting and design methodology for reduction of fuel, power and CO_2 on Total Sites, *Applied Thermal Engineering*, 17, 8–10, 993–1003.

Klemeš, J. and Perry, S.J. (2008). Methods to minimise energy use in food processing, in: Klemeš, J., Smith R. and Kim J.-K. (eds.), *Handbook of Water and Energy Management in Food Processing*, Cambridge, UK: Woodhead Publishing Limited, ISBN 1 84569 195 4; pp. 136–199.

Klemeš, J. and Ptáčník, R. (1985). Computer-aided synthesis of Heat Exchange Network, *Journal of Heat Recovery Systems*, 5 (5) 425–435.

Klemeš, J., Smith, R. and Kim, J.-K. (eds.) (2008). *Handbook of Water and Energy Management in Food Processing*, Cambridge, UK: Woodhead Publishing Limited, ISBN 1 84569 195 4, 1029 ps.

Klemeš, J. and Varbanov, P. (2010). Process Integration – Successful implementation and possible pitfalls, *Chemical Engineering Transactions*, 21, 1369–1374, DOI: 10.3303/CET1021229.

Klemeš, J.J, (2012), Industrial water recycle/reuse, *Current Opinion in Chemical Engineering*, doi: 10.1016/j.coche.2012.03.010

Klemeš, J.J. (ed.) (2013). *Process Integration Handbook*, Cambridge, United Kingdom: Woodhead Publishing, ISBN-13: 978 0 85709 593 0.

Klemeš, J.J., Friedler, F., Bulatov, I. and Varbanov, P.S. (2010). *Sustainability in the Process Industry – Integration and Optimization*, New York, USA: McGraw-Hill.

Linnhoff, B, (1972), *Thermodynamic Analysis of the Cement Burning Process* (Thermodynamische Analyse des Zementbrennprozesses), Diploma thesis, Abteilung IIIa, ETH Zurich (1972) (in German).

Linnhoff, B. (1979). *Thermodynamic Analysis in the Design of Process Networks*, PhD Thesis, The University of Leeds, Leeds, UK.

Linnhoff, B., Dunford, H. and Smith, R. (1983). Heat integration of distillation-columns into overall processes, *Chemical Engineering Science*, 38(8), 1175–1188, DOI:10.1016/0009–2509(83)80039–6.

Linnhoff, B. and Flower, J.R. (1978). Synthesis of heat exchanger networks: I. Systematic generation of energy optimal networks, *AIChE Journal*, 24(4), 633–642.

Linnhoff, B. and Hindmarsh, E. (1983). The pinch design method for heat exchanger networks, *Chem. Eng. Sci.* 38, 745–763.

Linnhoff, B. Mason, D.R. and Wardle, I. (1979). Understanding heat exchanger networks, *Computers and Chemical Engineering*, 3(1–4), 295–302.

Linnhoff, B., Townsend, D.W., Boland, D., Hewitt, G.F., Thomas, B.E.A., Guy, A.R. and Marsland, R.H. (1982). *A User Guide on Process Integration for the Efficient Use of Energy*, Rugby, UK: IChemE [revised edition published in 1994].

Manenti, F., Grottoli, M.G., and Pierucci, S., (2011), Online data reconciliation with poor-redundancy systems, *Industrial & Engineering Chemistry Research*, 50(24), 14105–14114

Mohammad Rozali, N.E., Wan Alwi, S.R., Manan, Z.A., Klemeš, J.J. and Hassan, M.Y. (2012). Process integration techniques for optimal design of hybrid power systems, *Applied Thermal Engineering*, doi: 10.1016/j.applthermaleng.2012.12.038.

Nishida, N., Liu, Y.A. and Lapidus, L. (1977). Studies in chemical process design and synthesis: III. A Simple and practical approach to the optimal synthesis of heat exchanger networks, *AIChE Journal*, 23(1), 77–93.

Pan, M., Bulatov, I., Smith, R. and Kim, J.-K. (2011). Novel optimization method for retrofitting heat exchanger networks with intensified heat transfer, *Computer Aided Chemical Engineering*, 29, 1864–1868.

Perry, S., Klemeš, J. and Bulatov, I. (2008). Integrating waste and renewable energy to reduce the carbon footprint of locally integrated energy sectors, *Energy*, 33 (10) 1489–1497.

PINET (2013). Process Intensification Network. www.pinetwork.org, accessed 13/06/2013.

Ponton, J.W. and Donaldson, R.A.B. (1974). A fast method for the synthesis of optimal heat exchanger networks, *Chemical Engineering Science*, 29, 2375–2377.

PRES – Process Integration, Modelling and Optimisation for Energy Saving and Pollution reduction (2013) www.confrence.pres.com accessed 08/06/2013.

PRES (2013). Conference Process Integration, Modelling and Optimisation for Energy Saving and Pollution Reduction. www.conferencepres.com, Accessed 31/03/2013.

Ptáčník, R. and Klemeš, J. (1987). Synthesis of optimum structure of Heat Exchange Networks, *Theoretical Foundations of Chemical Engineering*, 21(4), 488–499.

Ptáčník, R. and Klemeš, J. (1988). An application of mathematical methods in Heat-Exchange Network Synthesis, *Computers & Chemical Engineering*, 12(2/3), 231–235.

Reay, D. Ramshaw, C. and Harvey, A. (2008). *Process Intensification: Engineering for Efficiency, Sustainability and Flexibility*, Oxford, UK: Butterworth-Heinemann, ISBN-13: 978-0750689410.

Reay, D., Ramshaw, C. and Harvey, A. (2013). *Process Intensification, Engineering for Efficiency, Sustainability and Flexibility*, 2nd Edition, Oxford, UK: Butterworth-Heinemann, ISBN: 9780080983042, 512 pp.

Sargent, R.W.H. (1979). Flowsheeting, *Computers & Chemical Engineering*, 3(1–4), 17–20.

Sargent, R.W.H. (1983). Computers in chemical engineering – Challenges and constraints, *AIChE Symposium Series*, 79(235), 57–64.

Shelley, M.D. and El-Halwagi, M.M. (2000), Component-less design of recovery and allocation systems: A functionality-based clustering approach, *Computers and Chemical Engineering*, 24 (9–10), 2081–2091.

Smith, S. (2005). *Chemical Process: Design and Integration*, Chichester, UK: John Wiley & Sons Ltd.

Smith, R. (1995). *Chemical Process Design*, New York, USA: McGraw-Hill.

Smith, R., Klemeš, J., Tovazhnyansky, L.L., Kapustenko, P.A. and Uliev, L.M. (2000). *Foundations of Heat Processes Integration*, Kharkiv, Ukraine: NTU KhPI (in Russian).

Takama, N., Kuriyama, T., Shiroko, K. and Umeda, T. (1980). Optimal water allocation in a petroleum refinery, *Computers & Chemical Engineering*, 4(4), 251–258.

Tan, R. and Foo, D.C.Y. (2007). Pinch analysis approach to carbon-constrained energy sector planning, *Energy*, 32(8), 1422–1429.

Thevendiraraj, S., Klemeš, J., Paz, D., Aso, G. and Cardenas, G. J. (2003). Water and wastewater minimisation on a citrus plant, *Resources, Conservation and Recycling – International Journal of Sustainable Resource Management and Environmental Efficiency*, 37 (3), 227–250.

Townsend, D.W. and Linnhoff, B. (1983a). Heat and power networks in process design. Part I: Criteria for placement of heat engines and heat pumps in process networks, *American Institute of Chemical Engineering Journal*, 29(5), 742–748.

Townsend, D.W. and Linnhoff, B. (1983b). Heat and power networks in process design. Part II: Design procedure for equipment selection and process matching, *American Institute of Chemical Engineering Journal*, 29(5), 748–771.

Umeda, T., Harada, T. and Shiroko, K.A. (1979). Thermodynamic approach to the synthesis of heat integration systems in chemical processes, *Computers & Chemical Engineering*, 3(1–4), 273–282.

Umeda, T., Itoh, J. and Shiroko, K. (1978). Heat exchanger systems synthesis, *Chemical Engineering Progress* 74 (7) 70–76.

Wan Alwi, S.R. and Manan, Z. A. (2008). A holistic framework for design of cost-effective minimum water utilisation network, *Journal of Environmental Management*, 88, 219–252, DOI: 10.1016/j.jenvman.2007.02.011.

Wan Alwi, S.R. and Manan, Z.A. (2006). SHARPS – A new cost-screening technique to attain cost-effective minimum water utilisation network, *American Institute of Chemical Engineering Journal*, 11(52): 3981–3988.

Wan Alwi, S.R., Ismail, A., Manan, Z.A. and Handani, Z.B. (2011) A new graphical approach for simultaneous mass and energy minimization, *Applied Thermal Engineering*, 31 (6–7), 1021–1030.

Wan Alwi, S.R. and Manan, Z.A. (2010). STEP – A new graphical tool for simultaneous targeting and design of a heat exchanger network, *Chemical Engineering Journal*, 162 (1), 106–121.

Wan Alwi, S.R., Tin O.S., Mohammad Rozali, N.E., Abdul Manan, Z. and Klemeš, J.J. (2013). New graphical tools for process changes via load shifting for hybrid power systems based on Power Pinch Analysis, *Clean Technologies and Environmental Policy*, 15(3), 459–472, DOI: 10.1007/s10098-013-0605-7

Wang, Y.P. and Smith, R. (1994). Wastewater minimisation, *Chemical Engineering Science*, 49 (7) 981–1006.

Working Party of Computer Aided Process Engineering of European Federation of Chemical Engineering (2013). www.cape-wp.eu, accessed 01/04/2013.

Working Party on Process Intensification of European Federation of Chemical Engineering (2013). www.efce.info/Working+Parties/Process+Intensification/Home/Mission+Statement-p-111686.html, accessed 26/06/2013.

Zhu, X.X., Zanfir, M. and Klemeš J. (2000). Heat transfer enhancement for heat exchanger network retrofit, *Heat Transfer Engineering*, 21(2), 7–18.

2 Setting energy targets and Heat Integration

Heat Recovery can be used to provide either heating or cooling to processes to replace hot or cold utilities. It is widely applied in industry and has an extensive historical record. Systematic methods for performing Heat Recovery have emerged in the last 40 years inspired by the 1970s oil crises. Heat Recovery may take various forms: transferring heat between process streams, generating steam from higher temperature process waste heat, preheating a service stream (air for a furnace, as well as air or feed water for a boiler) by using excess process heat.

This chapter provides a systematic methodology for analysing the options for Heat Recovery in processes using limiting values of utility demands called targets. It also provides insights on how to combine one multi-stream process with energy intensive process units, or how to improve the processes internally using as a criterion the reduction of the utility targets. The main text is supplemented by working sessions for consolidating the knowledge and deepening the readers' understanding.

2.1 Introduction

Reducing the consumption of resources is typically achieved by increasing internal recycling and reuse of energy and material streams to replace the intake of fresh resources and utilities. Projects for improving process resource efficiency can offer economic benefits and also improve public perceptions of the company undertaking them. However, motivating, launching, and carrying out such projects requires a combination of several elements. First, as a necessary condition, come proper optimisation studies based on adequate models of the process plants. The sufficient condition is to provide an integral conceptual framework, where fundamentally sound concepts and efficient visualisation tools are used, combined with the optimisation. These together provide the plant and company managers with the necessary assurance and confidence in the proposed engineering solutions.

Process Integration (PI) is a family of methodologies for combining several processes to reduce consumption of resources and/or harmful emissions.

This chapter introduces the fundamentals of Heat Integration – the founding discipline of Process Integration. It discusses evaluation of Heat Recovery targets, Heat Exchange area targets, and basics of designing Heat Recovery networks. This Handbook should serve for the initial mastering of the methodology and for this reason some issues are simplified. However, they can be dealt with in detail when solving the real-life industrial problems. This includes e.g. fouling of heat exchangers (Gogenko et al., 2007).

2.1.1 Initial development of Heat Integration

The initial development and a brief history of Heat Integration – the first part of Process Integration – has been described in more details in the introductory Chapter 1. It is now over 40 years from the first appearance of the methodology until the most recent extensions and achievements. An overview can be found in Friedler (2009) and in more extended form in Friedler (2010). The methodology presented in this chapter has been also extended for batch processes. Some more information can be found elsewhere – e.g. Foo et al. (2008). A recommended case study has been published by Atkins et al. (2010).

2.1.2 Pinch Technology and targeting Heat Recovery: the thermodynamic roots

Klemeš et al.'s (2010) book includes a review of works tracing the development of the research on Heat Exchanger Networks. Their study shows that there was only mild interest in Heat Recovery and energy efficiency until the early 1970s, by which time just a few works in the field had appeared. But between the oil crises of 1973–1974 and 1979 there were significant advances made in Heat Integration. Although capital cost remained important, the major focus was on saving energy and reducing the related cost. It is exactly this focus that resulted in attention being paid to energy flows and to the energy quality represented by temperature. The result was the development of the Pinch Technology, which is firmly based on the first and second law of thermodynamics (Linnhoff and Flower, 1978).

In this way, Heat Exchanger Networks (HEN) synthesis – one of the very important and common tasks of process design – has become the starting point for the Process Integration revolution in industrial systems design. HEN in industry are used mainly to save energy costs. For many years the HEN design methods relied mostly on heuristics, as necessitated by the large number of permutations in which the necessary heat exchangers could be arranged. Masso and Rudd (1969) is a pioneering work that defines the problem of HEN synthesis. The paper proposes an evolutionary synthesis procedure based on heuristics. A complete timeline and thorough bibliography of HEN design and optimisation works is provided in Furman and Sahinidis (2002). The paper covers many more details, including the earliest known HEN-related scientific article: Ten Broeck (1944).

The discovery of the Heat Recovery Pinch concept (Linnhoff and Flower, 1978) was a critical step in the development of HEN synthesis. The main idea behind the formulated procedure was to obtain – prior to the core design steps – guidelines and targets for HEN performance. This procedure is possible thanks to thermodynamics. The hot and cold streams for the process under consideration are combined to yield (1) a Hot Composite Curve collectively representing the process heat sources (the hot streams); and (2) a Cold Composite Curve representing in a similar way the process heat sinks (the cold streams). For a specified minimum allowed temperature difference ΔT_{min}, the two curves are combined in one plot (Fig. 2.1), providing a clear thermodynamic view of the Heat Recovery problem.

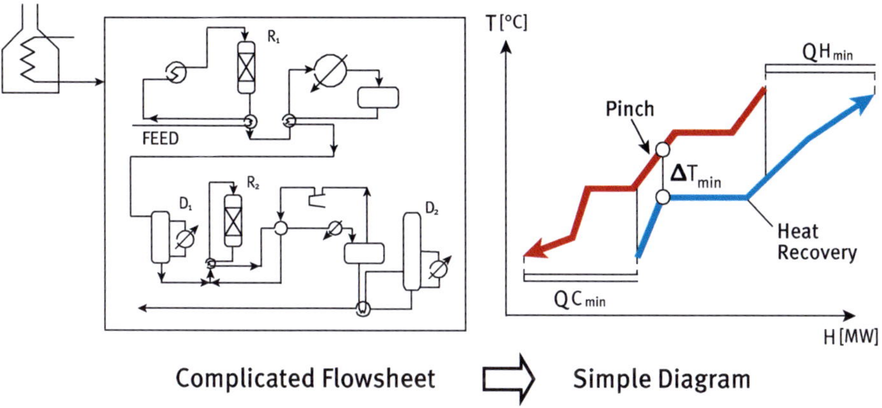

Fig. 2.1: Summary of Heat Recovery targeting.

The overlap between the two Composite Curves on the Heat Exchange axis represents the Heat Recovery target – i.e. the maximum amount of process heat being internally recovered. The vertical projection of the overlap indicates the temperature range where the maximum Heat Recovery should take place. The targets for external (utility) heating and cooling are represented by the non-overlapping segments of the Cold and Hot Composite Curves.

2.1.3 Supertargeting: full-fledged HEN targeting

After obtaining targets for utility demands of a Heat Exchanger Network, the next step is to estimate the targets for capital and total costs. The capital cost of a Heat Exchanger Network is determined by many factors. The most significant is the total heat transfer area and its distribution among the heat exchangers. Townsend and Linnhoff (1984) proposed a procedure for estimating HEN capital cost targets by using the Balanced Composite Curves, which are obtained by adding utilities to the Composite Curves obtained previously. Improvements to this procedure that have been proposed, and they involve one or more of the following factors:

1. Obtaining more accurate surface area targets for HENs that feature non-uniform heat transfer coefficients: Colberg and Morari (1990) use NLP transhipment models and account for forbidden matches; Jegede and Polley (1992) explored the distribution of the capital cost among different heat exchanger types; Zhu et al.

(1995) investigated integrated HEN targeting and synthesis within the context of the so-called "block decomposition" for HEN synthesis; non-uniform heat exchanger specifications; Serna-González et al. (2007) account also for different cost laws but do not put constraints on matches.

2. Accounting for practical implementation factors, such as construction materials, pressure ratings, and different heat exchanger types (Hall et al., 1990).

3. Accounting for additional constraints such as safety and prohibitive distance (Santos and Zemp, 2000).

2.1.4 Modifying the Pinch Idea for HEN retrofit

Bochenek et al. (1998) compared the approaches of optimisation versus simulation for retrofitting flexible HENs. This is an important work that should have generated additional research. Zhu et al. (2000) proposed a heat transfer enhancement methodology for HEN retrofit design, from which Heat Integration could benefit substantially. This approach is worthy of wider implementation, especially in the context of retrofit studies. There have been recent developments in this direction – for example Pan et al. (2013a) have developed an optimal HEN retrofit procedure for large-scale HENs, with a further follow-up focusing on shell-and-tube heat exchangers (Pan et al., 2013b).

Heat Exchanger Network retrofit is a special case of optimisation. In retrofit problems, an existing network with existing heat exchangers that had been already paid for has to be accommodated as much as possible. This requirement substantially alters the economics of the problem compared with developing a new design.

2.1.5 Benefits of Process Integration

The Composite Curves plot is a visual tool that summarises the important energy-related properties of a process in a single view (Fig. 2.1). It was the recognition of the thermodynamic relationships and limitations in the underlying Heat Recovery problem that led to development of the Pinch Design Method (Linnhoff and Hindmarsh, 1983). In addition, further applications have extended the PI approach to water minimisation as well as regional energy and emissions planning, financial planning, batch processes.

The most important property of thermodynamically derived Heat Recovery targets is that they cannot be improved upon by any real system. Composite Curves play an important role in process design. For HEN synthesis algorithms, they provide strict targets for maximum energy recovery. For process synthesis based on mathematical programming, the Composite Curves establish relevant lower bounds on utility requirements and capital cost, thereby narrowing the search space for the following superstructure construction and optimisation.

The preceding observation highlights an important characteristic of process optimisation problems, and specifically those that involve process synthesis and design. By strategically obtaining key data about the system, it is possible to evaluate processes based on limited information – before too much time (or other resources) are spent on more detailed analysis. This approach follows the logic of oil drilling projects: potential sites are first evaluated in terms of key preliminary indicators, and further studies or drilling commence only if the preliminary evaluations indicate that the revenues could justify further investment. The logic of this approach has been codified by Smith in his books on PI for process synthesis (Smith, 2005). Daichendt and Grossmann's (1997) paper integrates hierarchical decomposition and mathematical programming to solve process synthesis problems.

Process Integration has a direct impact on improving the sustainability of a given industrial process. All Process Intergration techniques are geared towards reducing the intake of resources and minimising the release of harmful effluents, goals that are directly related to the corresponding footprints. Hence, employing Process Integration and approaching the targeted values will help minimise those footprints.

2.2 Pinch Analysis for maximising energy efficiency

2.2.1 Introduction to Heat Exchange and Heat Recovery

Large amounts of thermal energy are used in industry to perform heating. Examples of this can be found in crude oil preheating before distillation, preheating of feed flows to chemical reactors, and heat addition to carry out endothermic chemical reactions. Some processes – such as condensation, exothermal chemical reactions, and product finalisation, similarly require heat to be extracted, which results in process cooling. There are several options for utility heating; these include steam, hot mineral oils, and direct fired heating. Steam is the most prevalent option because of its high specific heating value in the form of latent heat. Utility cooling options include water (used for moderate-temperature cooling when water is available), air (used when water is scarce or not economical to use), and refrigeration (when sub-ambient cooling is needed). Heat Recovery can be used to provide either heating or cooling to processes. Heat Recovery may take various forms: transferring heat between process streams, generating steam from higher temperature process waste heat, preheating (air for a furnace, as well as air or feed water for a boiler).

Heat transfer takes place in heat exchangers, which can employ either direct mixing or indirect heat transfer via a wall. Direct Heat Exchange is also referred to as non-isothermal mixing because the temperatures of the streams being mixed are different. Mixing heat exchangers are efficient at transferring heat and usually have low capital cost. In most industries, the bulk of the Heat Exchange must occur without mixing the heat-exchanging streams. In order to exchange only heat while keeping fluids separate, surface heat exchangers are employed. In these devices, heat is exchanged through a dividing wall. Because of its high thermal efficiency, the counter-current stream arrangement is the most common with surface heat exchangers. To simplify the discussion, counter-current heat exchangers are assumed unless stated otherwise. In terms of construction types, the traditional shell-and-tube heat exchanger is still the most common. However, plate-type heat exchangers are gaining increased attention (Godenko et al., 2007); their compactness, together with significant improvements in their resistance to leaking, have made them preferable in many cases.

2.2.1.1 Heat Exchange matches

A hot process stream can supply heat to a cold one when paired in one or several consecutive heat exchangers. Such pairing is referred to as a Heat Exchange match. The form of the steady-state balance equations for Heat Exchange matches that is most convenient for Heat Integration calculations is based on modelling a match as consisting of hot and cold sides, as shown in Fig. 2.2. The hot and cold part each have a simple, steady-state enthalpy balance that involves just one material stream and one heat transfer flow.

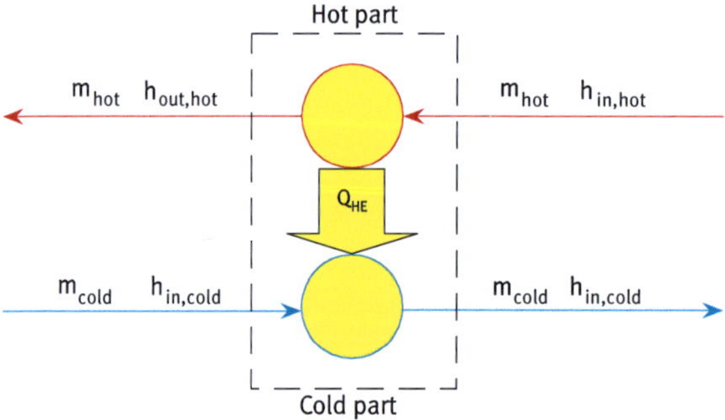

Fig. 2.2: Process Flow Diagram of a Heat Exchange match.

The main components of the model are (1) calculations of the heat transfer flows accounted for by the enthalpy balances and (2) estimation of the necessary heat transfer area. For the latter, both the log-mean temperature difference and the overall heat transfer coefficient are employed. The enthalpy balances of the hot and cold parts, and the kinetic equation of the heat transfer, may be written as follows:

$$Q_{HE} = m_{hot} \cdot (h_{in,\,hot} - h_{out,\,hot}) \tag{2.1}$$

$$Q_{HE} = m_{cold} \cdot (h_{out,\,cold} - h_{in,\,cold}) \tag{2.2}$$

$$Q_{\mathrm{HE}} = U \cdot A \cdot \Delta T_{\mathrm{LM}} \tag{2.3}$$

where Q_{HE} (kW) is the heat flow across the whole heat exchanger, U (kW·m^{-2}·°C^{-1}) is the overall heat transfer coefficient, A (m^2) is the heat transfer area, and ΔT_{LM} (°C) is the logarithmic-mean temperature difference. More information can be found in Shah and Sekulić (2003) – on fundamentals of heat exchanger design, Tovazshnyansky et al. (2004) – on industrial plate heat exchangers, and Shilling et al. (2008) containing general modelling and calculation.

2.2.1.2 Implementing Heat Exchange matches

The Heat Exchange matches are often viewed as being identical to heat exchangers, but this is not always the case. A given Heat Exchange match may be implemented by devices of different construction or by a combination of devices – for example, two counter-current heat exchangers in sequence may implement a single Heat Exchange match. The distinction between the concept of a Heat Exchange match and its implementation via heat exchangers is important because of capital cost considerations.

2.2.2 Basic principles

2.2.2.1 Process Integration and Heat Integration

At the time of the conception of Process Integration, attention was focused on reusing any waste heat generated on different sites. Each surface heat exchanger was described with only a few steady-state equations, and the thermal energy saved by

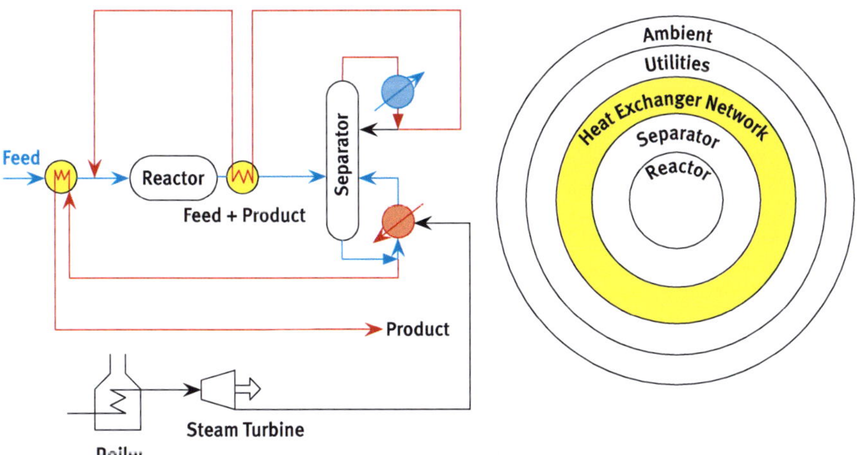

Fig. 2.3: The Onion Diagram.

reusing waste heat led to reductions in the expense of utility resources. This approach became popular under the names Heat Integration (HI) and Process Integration (PI). In this context, PI means integrating different processes to achieve energy savings. Engineers realised that integration could also reduce the consumption of other resources as well as the emission of pollutants. Heat and Process Integration came to be defined more widely in response to similar developments in water reuse and wastewater minimisation.

2.2.2.2 Hierarchy of process design

Process design has an inherent hierarchy that can be exploited for making design decisions. This hierarchy may be represented by the so-called Onion Diagram (Linnhoff et al., 1982, 1994), as shown in Fig. 2.3. The design of an industrial process starts with the reactors or other key operating units (the Onion's core). This is supplemented and served by other parts of the process, such as the separation subsystem (the next layer) and the Heat Exchanger Network (HEN) subsystem. The remaining heating and cooling duties, as well as the power demands, are handled by the utility system.

2.2.2.3 Performance targets

The thermodynamic bounds on Heat Exchange can be used to estimate the utility usage and Heat Exchange area for a given Heat Recovery problem. The resulting estimates of the process performance are lower bounds on the utility demands and a lower bound on the required heat transfer area. These bounds are known as targets for the reason that Heat Recovery estimates are achievable in practice and usually minimise the total cost of the HEN being designed.

2.2.2.4 Data extraction: Heat Recovery problem identification

For efficient Heat Recovery in industry, the relevant data must be identified and presented systematically. In the field of Heat Integration, this process is referred to as data extraction. The Heat Recovery problem data are extracted in several steps.
1. Inspect the general process flowsheet, which may contain Heat Recovery exchangers.
2. Remove the recovery heat exchangers and replace them with equivalent "virtual" heaters and coolers.
3. Lump all consecutive heaters and coolers.
4. The resulting virtual heaters and coolers represent the net heating and cooling demands of the flowsheet streams.
5. The heating and cooling demands of the flowsheet streams are then listed in a tabular format, where each heating demand is referred to as a cold stream and, conversely, each cooling demand as a hot stream.

This procedure is best illustrated by an example. Fig. 2.4 shows a process flowsheet involving two reactors and a distillation column. The process already incorporates two recovery heat exchangers. The utility heating demand of the process is Q_H = 1,760 kW, and the utility cooling demand is Q_C = 920 kW.

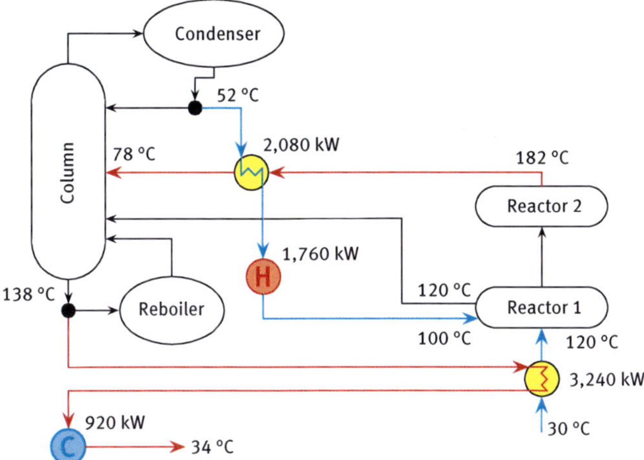

Fig. 2.4: Data extraction – An example process flowsheet.

The necessary thermal data have to be extracted from the initial flowsheet. Fig. 2.5 shows the flowsheet after steps 1 through 4. The heating and cooling demands of the streams have been consolidated by removing the existing exchangers. Reboiler and condenser duties have been left out of the analysis for simplicity (although these duties would be retained in an actual study). It is assumed that any process cooling duty is available to match up with any heating duty.

Applying step 5 to the data in Fig. 2.5 produces the data set in Table 2.1. By convention, heating duties are positive and cooling ones are negative. The subscripts S and T denote "supply" and "target" temperatures for the process streams.

Tab. 2.1: Data set for Heat Recovery Analysis.

Stream	Type	T_S (°C)	T_T (°C)	ΔH (kW)	CP (kW/°C)
H1	Hot	182	78	−2,080	20
H2	Hot	138	34	−4,160	40
C3	Cold	52	100	3,840	80
C4	Cold	30	120	3,240	36

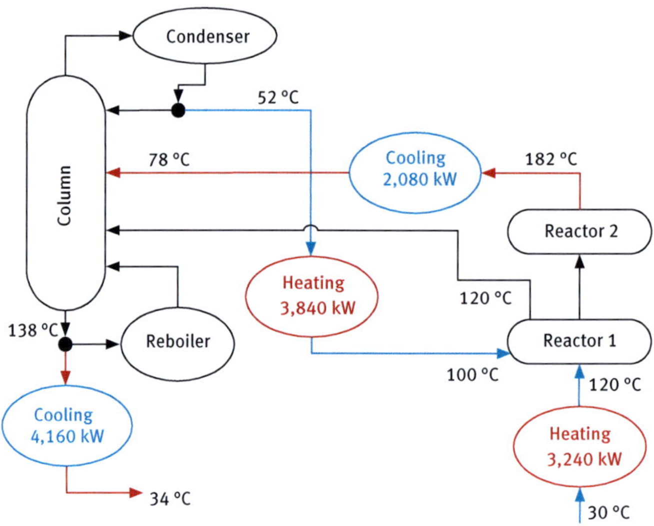

Fig. 2.5: Data extraction – heating and cooling demands.

The last column gives the heat capacity flow rate (CP). For streams that do not change phase (i.e. from liquid to gas or vice versa), CP is defined as the product of the specific heat capacity and the mass flow rate of the corresponding stream:

$$CP = m_{\text{stream}} \cdot C_{p,\,\text{stream}} \tag{2.4}$$

The CP can also be calculated using the following simple Equation (2.5):

$$CP = \frac{\Delta H}{T_T - T_S} \tag{2.5}$$

When phase transition occurs, the latent heat is used instead of CP to calculate the stream duties.

2.2.2.5 Working Session "Introduction to Heat Integration"
Assignment

Consider the flowsheet in Fig. 2.6. This is a modification of the example from the data extraction example in Fig. 2.5.

It features two reactors – R1 and R2 and a distillation column C1. They are connected by piping and the connections are characterised by temperature changes, necessitating heating and cooling. Focusing only on the heating and cooling requirements yields the entries presented in Table 2.2. This Heat Recovery problem comes with a constraint of minimum allowed temperature difference $\Delta T_{min} = 10\,°C$. Also given are two utilities:

- Steam (hot utility) at 200 °C
- Cooling water (CW, cold utility) at 25 °C.

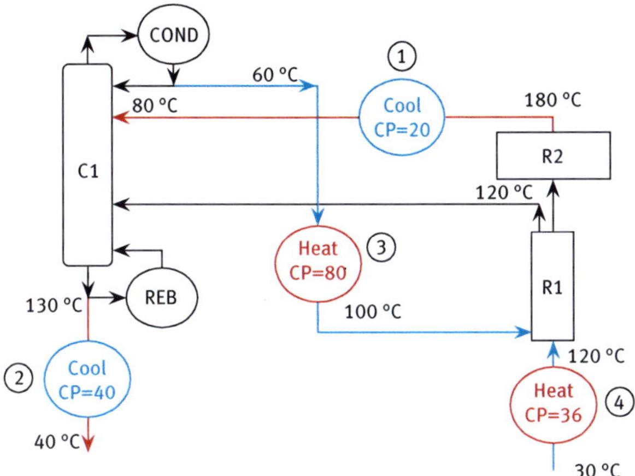

Fig. 2.6: Initial flowsheet for Working Session "Introduction to Heat Integration".

Tab. 2.2: Heat Recovery problem for Working Session "Introduction to Heat Integration".

Stream	T_S (°C)	T_T (°C)	CP (kW/°C)
1: Hot	180	80	20
2: Hot	130	40	40
3: Cold	60	100	80
4: Cold	30	120	36

The assignment is to design a network of steam heaters, water coolers and exchangers. Use Heat Recovery in preference to utilities.

Solution

Evaluation of heat exchanger placement options

For an ad hoc design of a Heat Exchanger Network, the first step is to construct a Cross-Grid Diagram as shown in Fig. 2.7, having the hot streams running from right to left and the cold streams from bottom to top. The grid features points where the hot and the cold streams intersect.

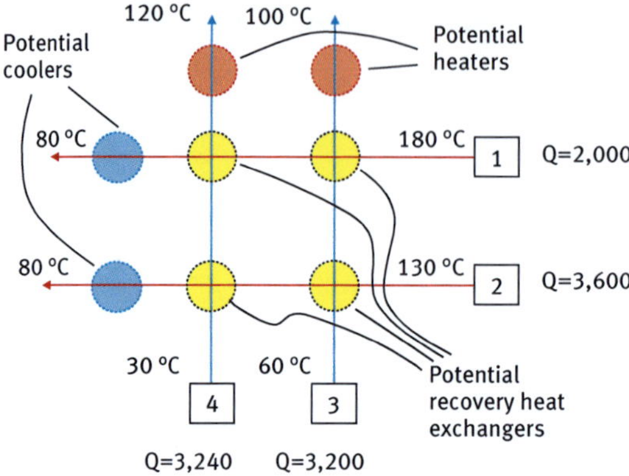

Fig. 2.7: Initial grid with placeholders for Working Session "Introduction to Heat Integration".

The next step is to evaluate the overall heating and cooling demands of the process streams:

$$1\,(\text{Hot}): \quad Q_1 = CP_1 \cdot (T_{S,1} - T_{T,1}) = 20 \cdot (180 - 80) = 2{,}000\,\text{kW} \tag{2.6}$$

$$2\,(\text{Hot}): \quad Q_2 = CP_2 \cdot (T_{S,2} - T_{T,2}) = 40 \cdot (130 - 40) = 3{,}600\,\text{kW} \tag{2.7}$$

$$3\,(\text{Cold}): \quad Q_3 = CP_3 \cdot (T_{T,3} - T_{S,3}) = 80 \cdot (100 - 60) = 3{,}200\,\text{kW} \tag{2.8}$$

$$4\,(\text{Cold}): \quad Q_4 = CP_4 \cdot (T_{T,4} - T_{S,4}) = 36 \cdot (120 - 30) = 3{,}240\,\text{kW} \tag{2.9}$$

Further, one can evaluate the options for placing heat exchangers. Cooling down Stream 1 (hot) by matching with Stream 3 (cold) is evaluated in Fig. 2.8.

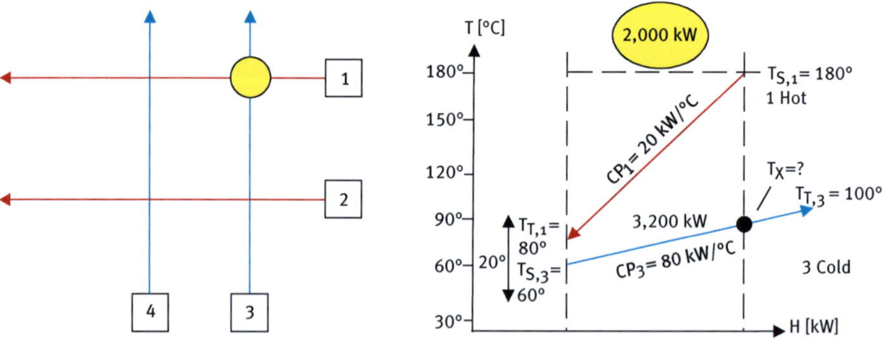

Fig. 2.8: Option 1 – Matching Streams 1 and 3 (Working Session "Introduction to Heat Integration").

It can be noticed that the smaller temperature difference for this potential match is at the cold end and it is 20 °C. In the direction of increasing the temperatures, the two T-H profiles diverge, which indicates that such a match is feasible. Stream 1 has the smaller load (2,000 kW). This is, therefore, selected as the heat exchanger duty: Q = 2,000 kW. Such a selection means that the final temperature of the hot stream of 80 °C will be reached and its cooling requirement will be completely satisfied.

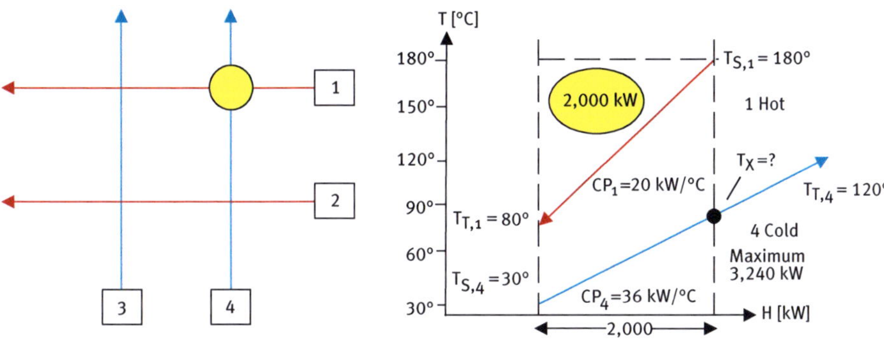

Fig. 2.9: Option 2 – Matching Streams 1 and 4 (Working Session "Introduction to Heat Integration").

The exit temperature T_X of Stream 3 will be lower than the final one and the heating requirement of Stream 3 will be satisfied only partially. The temperature T_X (from Fig. 2.8) can be calculated as follows:

$$T_X = T_{S,3} + \frac{Q}{CP_3} = 60 + \frac{2,000}{80} = 60 + 25 = 85\,°C. \tag{2.10}$$

Option 2 involves matching Streams 1 (hot) and 4 (cold) – Fig. 2.9. In a similar analysis as for Option 1, the heat exchanger duty would be 2,000 kW and Stream 4 will be satisfied only partially, reaching temperature 85.6 °C instead of the desired 120 °C.

The third option is to place a heat exchanger match between Streams 2 (hot) and 3 (cold) as shown in Fig. 2.10. Since Stream 2 (hot) has a smaller CP value than that for Stream 3 (cold), when starting to follow the profiles from right to left (from higher temperatures to lower), the T-H profiles of the two streams converge. This indicates that starting from a temperature difference of 30 °C, inside the heat exchanger the temperature difference would become smaller and smaller, passing the lower bound of $\Delta T_{min} = 10\,°C$ and eventually zero. This is an infeasible condition and such an option cannot be used.

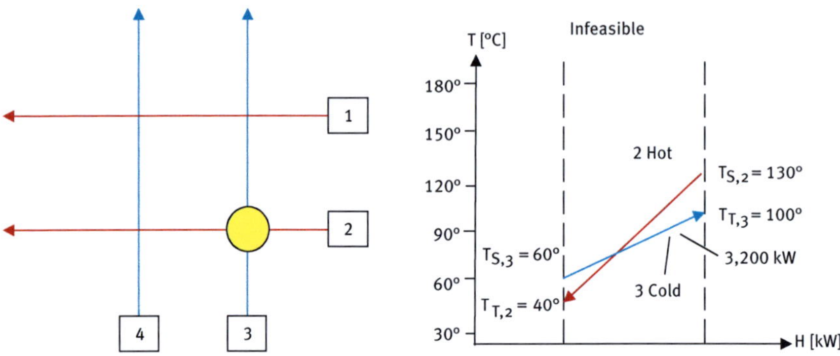

Fig. 2.10: Option 3 – Matching Streams 2 and 3 (Working Session "Introduction to Heat Integration").

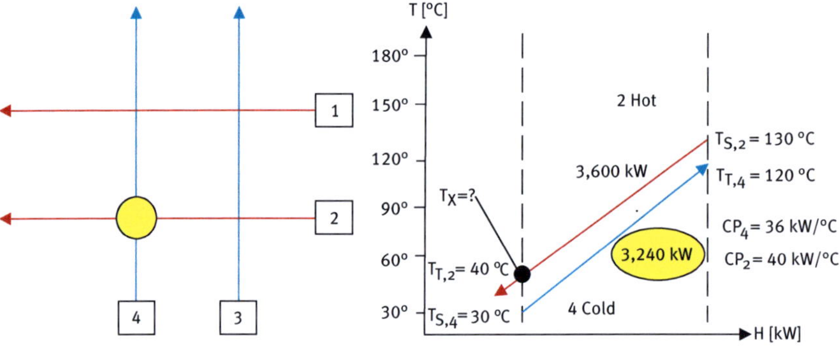

Fig. 2.11: Option 4 – Matching Streams 2 and 4 (Working Session "Introduction to Heat Integration").

Option 4 provides the final slot for a match – to match Streams 2 (hot) and 4 (cold) – see Fig. 2.11. Stream 4 has the smaller load (3,240 kW) and this is selected as the heat exchanger duty: Q = 3,240 kW.

- The final temperature of Stream 4 (120 °C) will be reached.
- The exit temperature T_x of Stream 2 will be higher than the final one.

$$T_X = T_{S,2} - \frac{Q}{CP_2} = 130 - \frac{3240}{40} = 130 - 81 = 49\,°C. \tag{2.11}$$

Possible solutions

One possible solution is to select Options 1 and 4 for recovery Heat Exchange and satisfy the remaining heating requirement with a heater and the remaining cooling requirement with a cooler. The resulting network is shown in Fig. 2.12. The network in Fig. 2.12 is not the only solution. A different network allowing maximum Heat Integration is shown in Fig. 2.13.

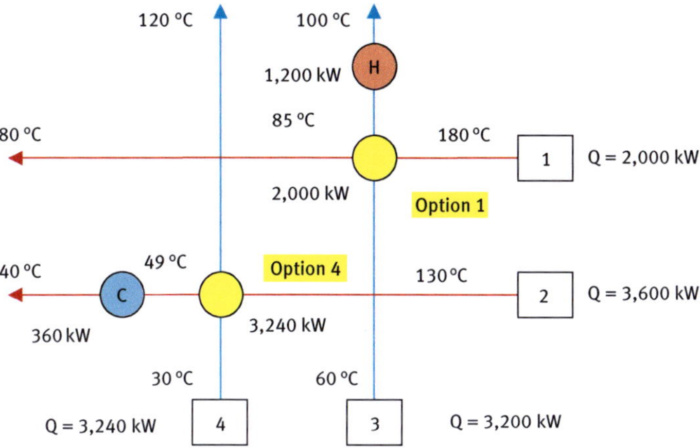

Fig. 2.12: One possible HEN (Working Session "Introduction to Heat Integration").

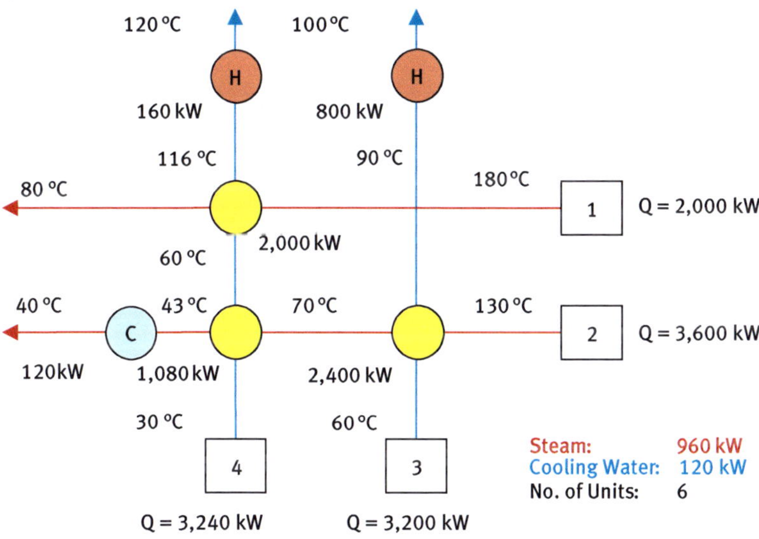

Fig. 2.13: HEN with maximum Heat Integration (Working Session "Introduction to Heat Integration").

Another network, featuring a minimum number of heat exchanger units, can be seen in Fig. 2.14.

Another interesting solution can be obtained if the minimum allowed temperature difference is reduced to $\Delta T_{min} = 6.6\,°C$. In this case the maximum Heat Recovery eliminates the need for cooling water. The network of this solution is shown in Fig. 2.15.

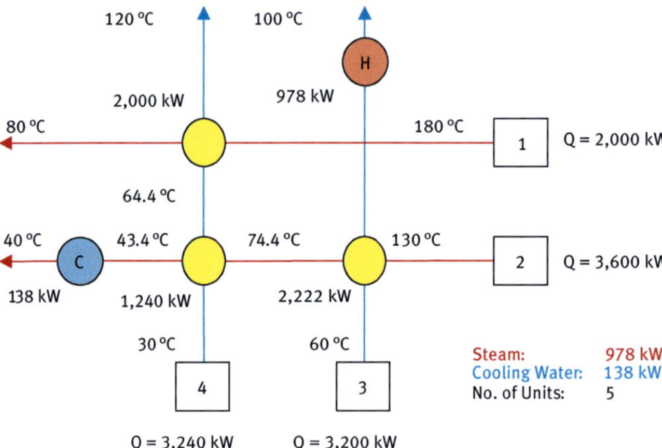

Fig. 2.14: HEN featuring minimum number of units (Working Session "Introduction to Heat Integration").

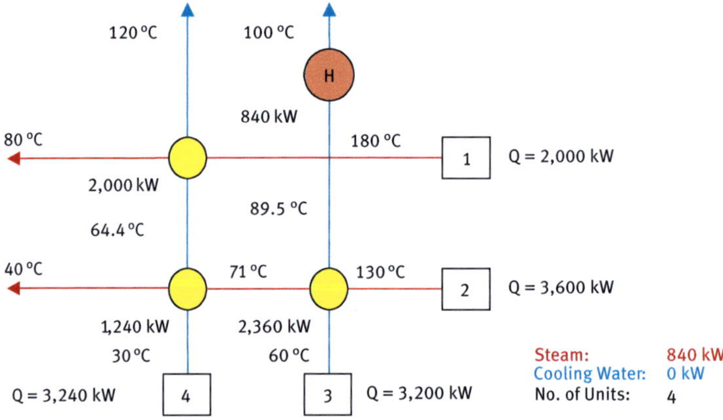

Fig. 2.15: HEN eliminating the need for cooling water (Working Session "Introduction to Heat Integration").

2.2.3 Basic Pinch Technology

The main strategy of Pinch-based Process Integration is to identify the performance targets before starting the core process design activity. Following this strategy yields important clues and design guidelines. The most common hot utility is steam. Heating with steam is usually approximated as a constant-temperature heating utility. Cooling with water is non-isothermal for the carrier because the cooling effect results from sensible heat absorption into the water stream and thus leads to increasing its temperature.

2.2.3.1 Setting energy targets

Heat Recovery between one hot and one cold stream

The second law of thermodynamics implies that heat flows from higher temperature to lower temperature locations. As shown in Equation (2.3), in a heat exchanger the required heat transfer area is proportional to the temperature difference between the streams.

In heat exchanger design, the minimum allowed temperature difference (ΔT_{min}) is the lower bound on any temperature differences to be encountered in any heat exchanger in the network. The value of ΔT_{min} is a design parameter determined by exploring the trade-offs between more Heat Recovery and the larger heat transfer area requirement. Any given pair of hot and cold process streams may exchange as much heat as allowed by their temperatures and the minimum temperature difference.

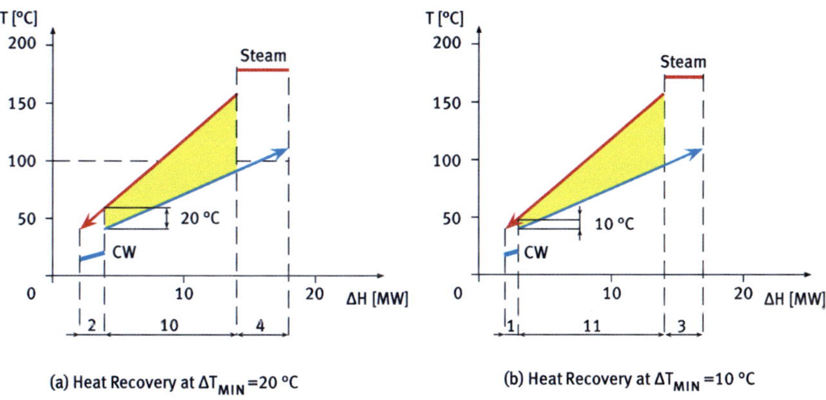

Fig. 2.16: Thermodynamic limits on Heat Recovery.

Consider the two-stream example shown in Fig. 2.16(a). The amount of Heat Recovery is 10 MW, which is achieved by allowing $\Delta T_{min} = 20\ °C$. If $\Delta T_{min} = 10\ °C$, as in Fig. 2.6(b), then it is possible to "squeeze out" one more MW of Heat Recovery. For each value of ΔT_{min} it is possible to identify the maximum possible Heat Recovery between two process streams. To obtain the Heat Recovery targets for a HEN design problem, this principle needs to be extended to handle multiple streams.

Evaluation of Heat Recovery for multiple streams: the Composite Curves

The analysis starts by combining all hot streams and all cold streams into two Composite Curves – CCs (Linnhoff et al., 1982). For each process there are two curves: one for the hot streams (Hot Composite Curve – HCC) and another for the cold streams (Cold Composite Curve – CCC). Each CC consists of a temperature-enthalpy (T-H) profile, representing the overall heat availability in the process (the HCC) and the overall heat demands of the process (the CCC). The procedure of HCC construction is illustrated in Fig. 2.17 on the data from Table 2.1.

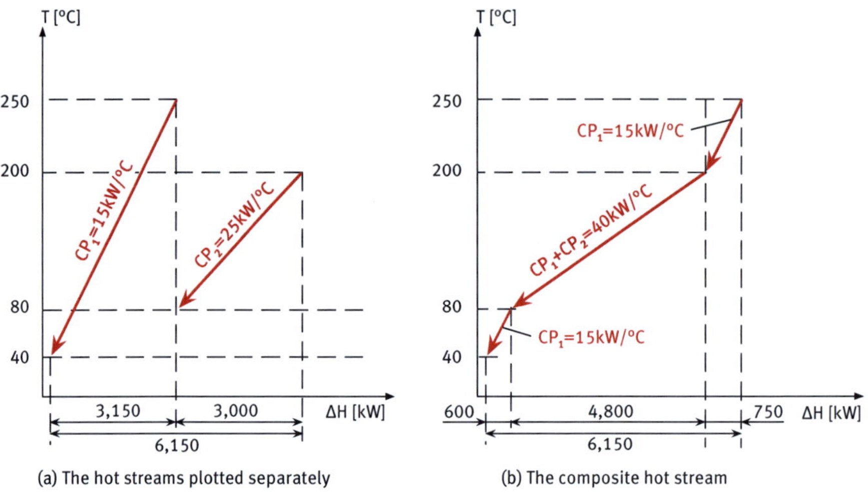

(a) The hot streams plotted separately   (b) The composite hot stream

Fig. 2.17: Constructing the Hot Composite Curve (HCC).

All temperature intervals are formed by the starting and target temperatures of the hot process streams. Within each temperature interval, a composite segment is formed consisting of (1) a temperature difference equal to that of the interval and (2) a total cooling requirement equal to the sum of the cooling requirements of all streams within the interval by summing up the heat capacity flow rates of the streams. The composite segments from all temperature intervals are combined to form the HCC. Construction of the Cold Composite Curve is entirely analogous.

The two Composite Curves are combined in the same plot in order to identify the maximum overlap, which represents the maximum amount of heat that could be recovered. The HCC and CCC for the example from Table 2.1 are shown together in Fig. 2.18.

Both CCs can be moved horizontally (i.e., along the ΔH axis), but usually the HCC position is fixed and the CCC is shifted. This is equivalent to varying the amount of Heat Recovery and (simultaneously) the amount of required utility heating and cooling. Where the curves overlap, heat can be recovered between the hot and cold streams. More overlap means more Heat Recovery and smaller utility requirements, and vice versa. As the overlap increases, the temperature differences between the overlapping curve segments decrease. Finally, at a certain overlap, the curves reach the minimum allowed temperature difference, ΔT_{min}. Beyond this point, no further overlap is possible. The closest approach between the curves is termed the *Pinch Point* (or simply the *Pinch*); it is also known as the *Heat Recovery Pinch*.

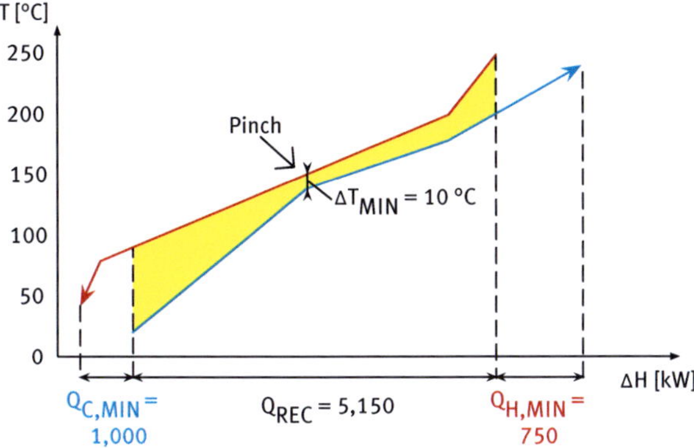

Fig. 2.18: The HCC and CCC at $\Delta T_{min} = 10\ °C$.

It is important to note that the amount of largest overlap (and thus the maximum Heat Recovery) would be different if the minimum allowed temperature difference is changed for the same set of hot and cold streams. The larger the value of ΔT_{min}, the smaller the possible maximum Heat Recovery. Specifying the minimum utility heating, the minimum utility cooling, or the minimum temperature difference fixes the relative position of the two Composite Curves and hence the maximum possible amount of Heat Recovery. The identified Heat Recovery targets are relative to the specified value of ΔT_{min}. If that value is increased, then the minimum utility requirements also increase and the potential for maximum recovery drops (Fig. 2.19).

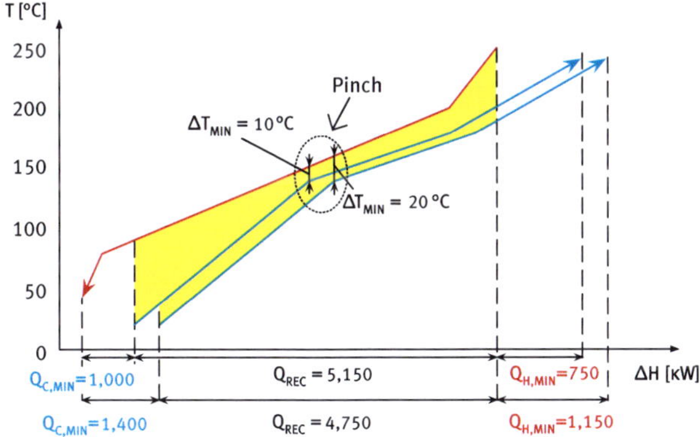

Fig. 2.19: Variation of Heat Recovery targets with ΔT_{min}.

The appropriate value for ΔT_{min} is determined by economic trade-offs. Increasing ΔT_{min} results in larger minimum utility demands and increased energy costs; choosing a higher value reflects the need to reduce heat transfer area and its corresponding investment cost. Conversely, if ΔT_{min} is reduced then utility costs go down but investment costs go up. This trade-off is illustrated in Fig. 2.20.

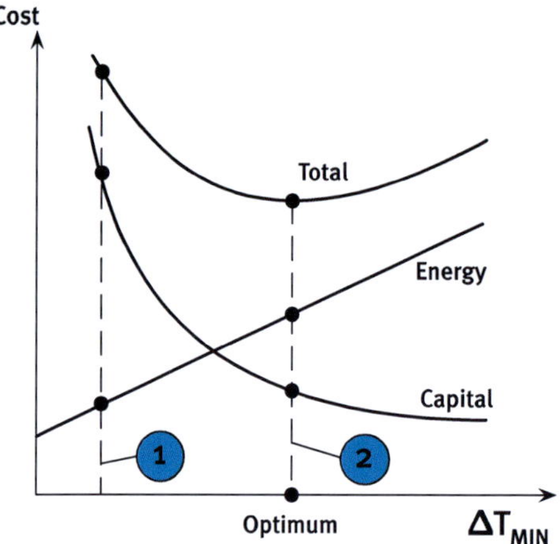

Fig. 2.20: Trade-off between investment and energy costs as a function of ΔT_{min}.

2.2.3.2 Working Session "Setting energy targets"
Assignment

Consider again the flowsheet from Fig. 2.6. As already discussed, applying data extraction to that produces the stream data set from Table 2.2. To refresh, the minimum allowed temperature difference is $\Delta T_{min} = 10\,°C$ and also given are two utilities:
- Steam (hot utility) at 200 °C
- CW (cold utility) at 25 °C.

The assignment is to construct the Composite Curves and then to identify the utility targets.

Solution

First the Hot Composite Curve is constructed. This starts with plotting the hot streams separately as shown in Fig. 2.21.

The starting and ending temperatures of the hot streams are listed and sorted in descending order (Fig. 2.22). Afterwards, the sorted list is used to form temperature intervals and calculate the enthalpy balances (Fig. 2.23).

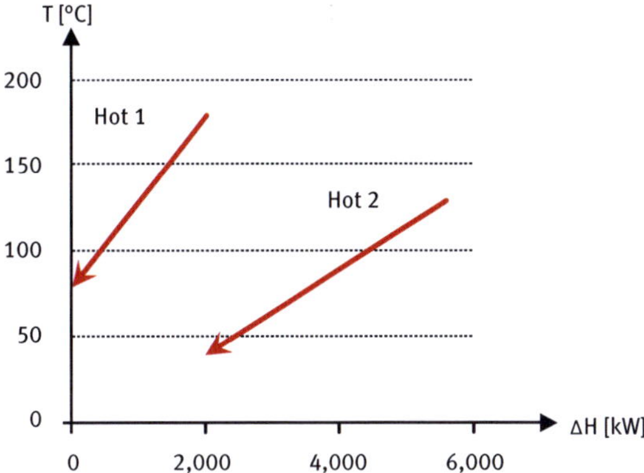

Fig. 2.21: The hot streams for Working Session "Setting energy targets" plotted separately.

Initial list

Stream	T [°C]
Hot 1	180
Hot 1	80
Hot 2	130
Hot 2	40

Sorted

Stream	T [°C]
Hot 1	180
Hot 2	130
Hot 1	80
Hot 2	40

Fig. 2.22: Starting and ending temperatures of the hot streams for Working Session "Setting energy targets".

	$\Sigma CP_{interval}$	Enthalpy Balances
Hot 1 $CP = 20$ kW/ °C **180 °C**		
Hot 2 $CP = 40$ kW/ °C	$\Sigma CP_{interval} = 20$ kW/ °C (only Hot 1 present)	$= \Sigma CP_{interval} \cdot \Delta T_{interval} = $ $= 20 \cdot (180\text{-}130) = 1{,}000$ kW
130 °C	$\Sigma CP_{interval} = 40+20 = $ $= 60$ kW/ °C (Hot 1 and Hot 2)	$= 60 \cdot (130\text{-}80) = 3{,}000$ kW
80 °C	$\Sigma CP_{interval} = 40$ kW/ °C (only Hot 2 present)	$= 40 \cdot (80\text{-}40) = 1{,}600$ kW
40 °C		

Fig. 2.23: Enthalpy balances for combining the hot streams for Working Session "Setting energy targets".

The Hot Composite Curve is plotted using the obtained enthalpy balances by starting from enthalpy change zero and the lowest interval temperature (40 °C), then moving upwards in the table from Fig. 2.23. The resulting plot is shown in Fig. 2.24.

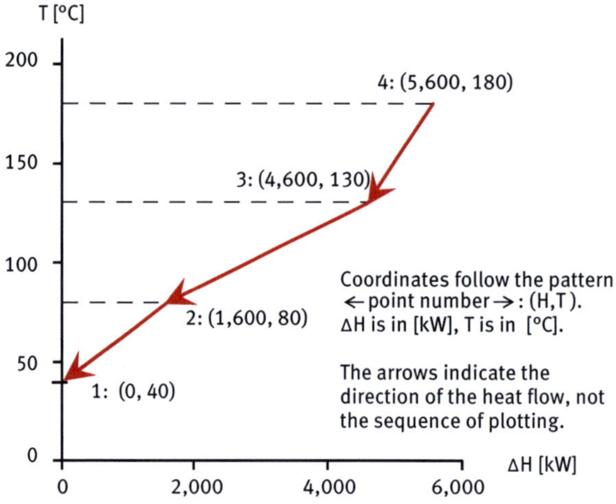

Fig. 2.24: The Hot Composite Curve for Working Session "Setting energy targets".

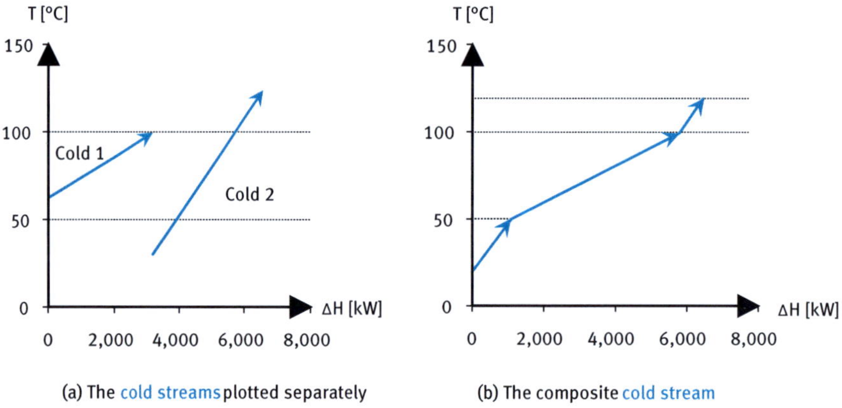

(a) The cold streams plotted separately

(b) The composite cold stream

Fig. 2.25: The Cold Composite Curve for Working Session "Setting energy targets".

The Cold Composite Curve is constructed in a similar way (illustrated in Fig. 2.25).

The two Composite Curves are put together obtaining the plot in Fig. 2.26. The resulting targets are:
- Pinch location at 70/60 °C
- Minimum utility heating 960 kW
- Minimum utility cooling 120 kW.

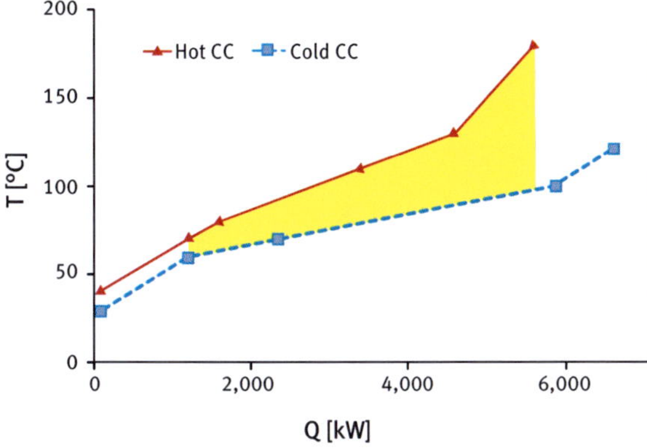

Fig. 2.26: The two Composite Curves for Working Session "Setting energy targets".

2.2.3.3 The Heat Recovery Pinch

The Heat Recovery Pinch has important implications on the HEN being designed. As illustrated in Fig. 2.27, the Pinch sets the absolute limits for Heat Recovery within the process.

The Pinch point divides the Heat Recovery problem into a net heat sink above the Pinch point and a net heat source below it (Fig. 2.28). At the Pinch point, the temperature difference between the hot and cold streams is exactly equal to ΔT_{min}, which means that at this point the streams are not allowed to exchange heat. As a result, the heat sink above the Pinch is in balance with the minimum hot utility ($Q_{H,min}$) and the heat source below the Pinch is in balance with the minimum cold utility ($Q_{C,min}$); thus, no heat is transferred across the Pinch via utilities or via process-to-process heat transfer.

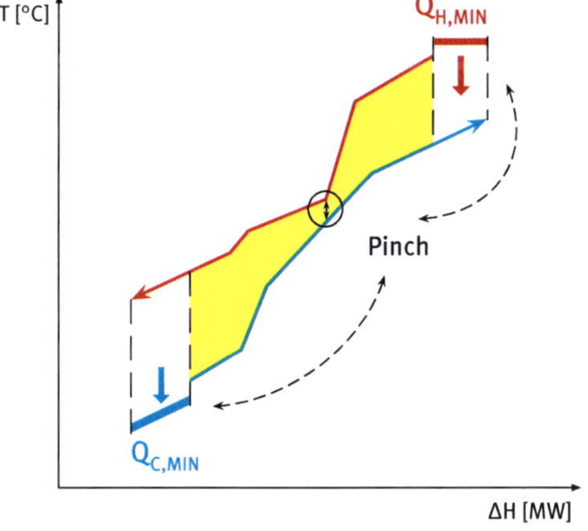

Fig. 2.27: Limits for process Heat Recovery set by the Pinch.

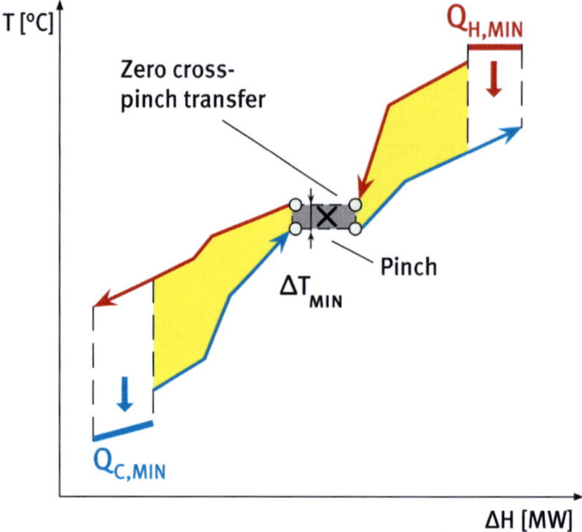

Fig. 2.28: Partitioning the Heat Recovery problem.

No heat can be transferred from below to above the Pinch, because this is thermodynamically infeasible. However, it is feasible to transfer heat from hot streams above the Pinch to cold streams below the Pinch. All cold streams – even those below the Pinch – could be heated by a hot utility. Likewise, the hot streams (even above the Pinch) could be cooled by a cold utility. Although these arrangements are thermodynamically feasible, applying them would cause utility use to exceed the minimum, as identified by the Pinch Analysis. This is why the Pinch is fundamental to the design of Heat Recovery systems.

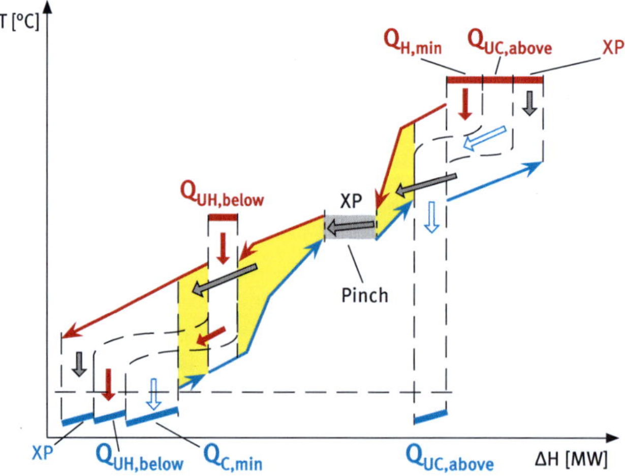

Fig. 2.29: More in – more out.

It is important to discuss what happens if heat is transferred across the Pinch. Recall that it is possible to transfer heat only from above to below the Pinch. If, say, XP units of heat are transferred across the Pinch (Fig. 2.29), then $Q_{H,min}$ and $Q_{C,min}$ will each increase by the same amount in order to maintain the heat balances of the two problem parts. Any extra heat that is added to the system by the hot utility has to be taken away by the cold utility, in addition to the minimum requirement $Q_{C,min}$.

Cross-Pinch process-to-process heat transfer is not the only way by which a problem's thermodynamic Pinch partitioning can be violated. This could also happen if the external utilities are placed incorrectly. For example, any heating below the Pinch will create a need for additional cooling in that part of the system (Fig. 2.29). Conversely, any utility cooling above the Pinch will create a need for additional utility heating. The implications of the Pinch for Heat Recovery problems can be distilled into the following three conditions, which have to hold if the minimum energy targets for a process are to be achieved:

1. Heat must not be transferred across the Pinch.
2. There must be no external cooling above the Pinch.
3. There must be no external heating below the Pinch.

Violating any of these rules will lead to an increase in energy utility demands. The rules are applied explicitly in the context of HEN synthesis by the Pinch Design Method (Linnhoff and Hindmarsh, 1983) and also before starting a HEN retrofit analysis to identify causes of excessive utility demands by a process. Other HEN synthesis methods – if they achieve the minimum utility demands – also conform to the Pinch rules (though sometimes only implicitly).

2.2.3.4 Numerical targeting: The Problem Table Algorithm

The Composite Curves are a useful tool for visualising Heat Recovery targets. However, they can be time consuming to draw for problems that involve many process streams. In addition, targeting that relies solely on such graphical techniques cannot be very precise.

Tab. 2.3: PTA example: Process streams data ($\Delta T_{min} = 10°C$).

No.	Type	T_S (°C)	T_T (°C)	CP (kW/°C)	T_S* (°C)	T_T* (°C)
1	Cold	20	180	20	25	185
2	Hot	250	40	15	245	35
3	Cold	140	230	30	145	235
4	Hot	200	80	25	195	75

The process of identifying numerical targets is therefore usually based on an algorithm known as the Problem Table Algorithm (PTA). Some authors employ the equivalent "transshipment" model (Cerdá et al., 1990).

The steps are as follows:

1. Shift the process stream temperatures.
2. Set up temperature intervals.
3. Calculate interval heat balances.
4. Assuming zero hot utility, cascade the balances as heat flows.
5. Ensure positive heat flows by increasing the hot utility as needed.

The algorithm will be illustrated using the sample data in Table 2.3.

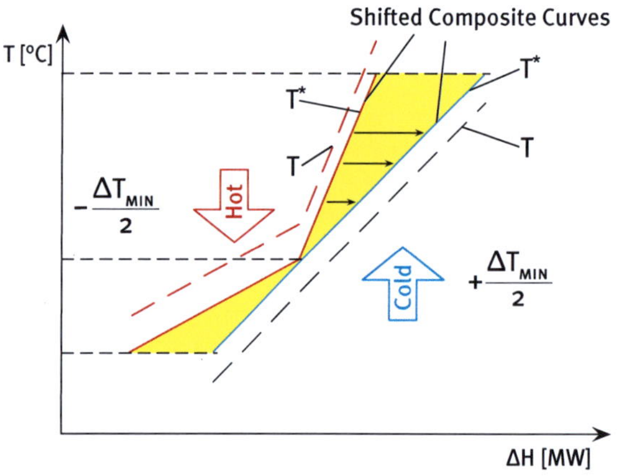

Fig. 2.30: Temperature shifting to ensure feasible heat transfer.

Step 1: Because the PTA uses temperature intervals, it is necessary to set up a unified temperature scale for the calculations. If the real stream temperatures are used, then some of the heat content would be left out of the recovery. The problem is avoided by obtaining shifted stream temperatures (T^*) for PTA calculations. Thus, the hot streams are shifted to be colder by $\Delta T_{min}/2$ and the cold streams are shifted to be hotter by $\Delta T_{min}/2$. Hence, if the shifted temperatures (T^*) of a cold and a hot stream (or their parts) are the same, then their real temperatures are still actually ΔT_{min} apart, which allows for feasible heat transfer. This operation is equivalent to shifting the Composite Curves toward each other vertically, as illustrated in Fig. 2.30. The last two (shaded) columns in Table 2.3 show the shifted process stream temperatures.

Step 2: Temperature intervals are formed by listing all shifted process stream temperatures in descending order (any duplicate values are considered just once). This action creates temperature boundaries (TBs), which form the temperature intervals for the problem. For the example in Table 2.3, the TBs are 245 °C, 235 °C, 195 °C, 185 °C, 145 °C, 75 °C, 35 °C, and 25 °C.

Step 3: The heat balance is calculated in each temperature interval. First, the stream population of the process segments falling within each temperature interval (the first two columns of Table 2.4) is identified. The sums of the segment CPs (heat capacity flow rates) in each interval are calculated; then that sum is multiplied by the interval temperature difference (i.e., the difference between the TBs that define each interval). This calculation is also illustrated in Table 2.4.

Tab. 2.4: Problem Table Algorithm (PTA) for the streams in Table 2.3.

Interval Temp [°C]	Stream Population	$\Delta T_{\text{INTREVAL}}$ [°C]	$\Sigma CP_H - \Sigma CP_C$ [kW/°C]	$\Delta H_{\text{INTREVAL}}$ [kW]	Surplus/ Deficit
245	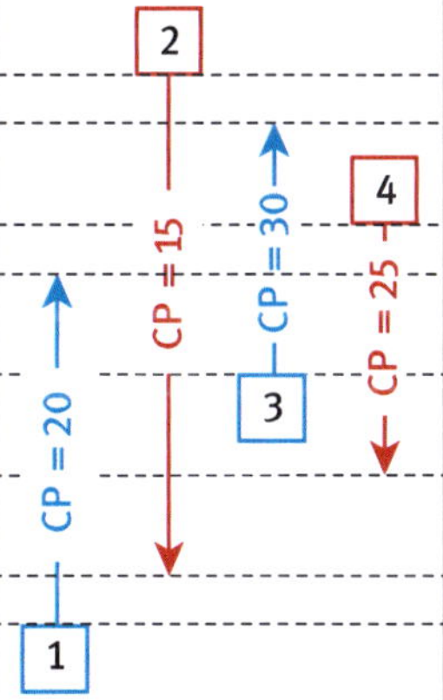				
235		10	15	150	Surplus
		40	−15	−600	Deficit
195		10	10	100	Surplus
185		40	−10	−400	Deficit
145		70	20	1400	Surplus
75		40	−5	−200	Deficit
35		10	20	−200	Deficit
25					

Step 4: The Problem Heat Cascade shown in Fig. 2.31 has a box allocated to each temperature interval; each box contains the corresponding interval enthalpy balances. The boxes are connected with heat flow arrows in order of descending temperature. The top heat flow represents the total hot utility provided to the cascade, and the bottom heat flow represents the total cold utility. The hot utility flow is initially assumed to be zero and this value is combined (summed up) with the enthalpy balance of the top cascade interval to produce the value for the next lower cascade heat flow. This operation is repeated for the lower temperature intervals and connecting heat flows until the bottom heat flow is calculated, resulting in the cascade shown in Fig. 2.31(a).

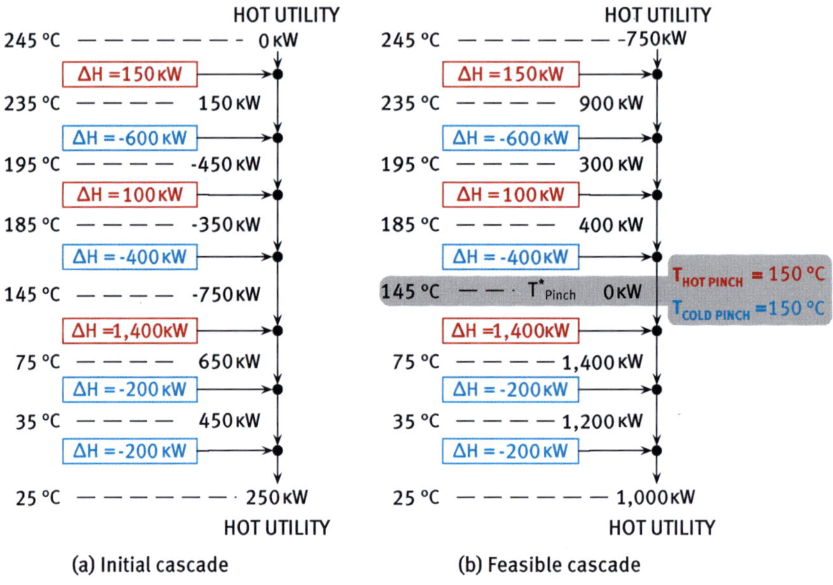

Fig. 2.31: Heat cascade for the process data in Table 2.3.

Step 5: The resulting heat flow values in the cascade are examined, and a feasible heat cascade is obtained; see Fig. 2.31(b). From the cascading heat flows, the smallest value is identified; if it is nonnegative (i.e., positive or zero), then the heat cascade is thermodynamically feasible. If a negative value is obtained, then a positive utility flow of the same absolute value has to be provided at the topmost heat flow, after which the cascading described in Step 4 is repeated. The resulting heat cascade is guaranteed to be feasible and provides numerical Heat Recovery targets for the problem. The topmost heat flow represents the minimum hot utility, the bottommost heat flow represents the minimum cold utility, and the TB with zero heat flow represents the location of the (Heat Recovery) Pinch. It is often possible to obtain more than one zero-flow temperature boundary, each representing a separate Pinch point.

2.2.3.5 Working Session "The Problem Table Algorithm"
Assignment

The same flowsheet as before is considered (Fig. 2.6). The process stream data is the same (Table 2.2), so are the remaining specifications of the Heat Recovery problem (ΔT_{min} = 10 °C, steam at 200 °C, CW at 25 °C). The task is, using the Problem Table Algorithm, to calculate the minimum hot and cold utility requirements and the location of the Pinch.

Solution

First the stream temperatures are shifted by $\Delta T_{min}/2$ – hot streams down and cold streams up, as shown in Table 2.5.

Tab. 2.5: Obtaining the shifted temperatures (Working Session "The Problem Table Algorithm").

Stream	T_S, (°C)	T_T, (°C)	T_S^* (°C)	T_T^*, (°C)
1 (hot)	180	80	175	75
2 (hot)	130	40	125	35
3 (cold)	60	100	65	105
4 (cold)	30	120	35	125

After eliminating the duplicates and sorting in descending order, the following list of temperature boundaries remains: 175, 125, 105, 75, 65, 35 °C. This allows to further draw the process streams population against the temperature boundaries, to form the temperature intervals and to calculate the enthalpy balances of the intervals (Table 2.6).

Tab. 2.6: The Problem Table for Working Session "The Problem Table Algorithm".

Shifted Interval Temperature	Stream Population	$\Delta T_{interval}$	$\Sigma CP_H - \Sigma CP_C$	$\Delta H_{interval}$	Surplus/ Deficit
175					
		50	20	1,000	Surplus
125					
		20	24	480	Surplus
105					
		30	−56	−1,680	Deficit
75					
		10	−76	−760	Deficit
65					
		30	4	120	Surplus
35					

Stream population CP values: Stream 1 CP = 20, Stream 2 CP = 40, Stream 3 CP = 80, Stream 4 CP = 36.

The enthalpy balances of the temperature intervals are used to construct the heat cascade. At the first stage a zero hot utility is assumed and the balances propagated (Fig. 2.32(a)) and then the negative flows are eliminated to obtain the feasible heat cascade (Fig. 2.32(b)).

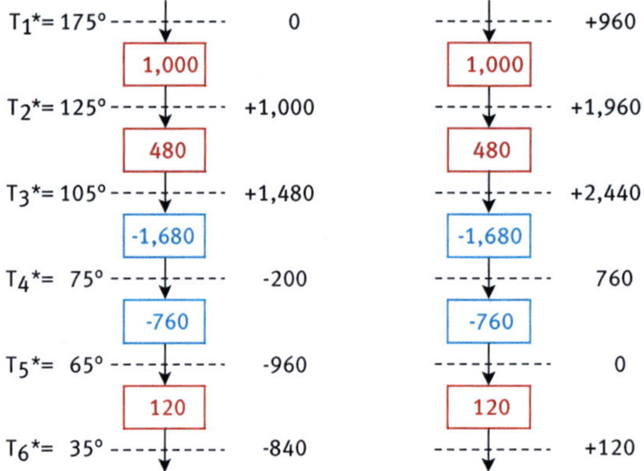

Fig. 2.32: Heat cascade for Working Session "The Problem Table Algorithm".

In the feasible cascade (Fig. 2.32(b)), the temperature boundary with zero heat flow is at 65 °C, which indicates the Pinch location as 70/60 °C. The minimum utility heating is 960 kW identified by the heat flow at the top of the cascade and the minimum utility cooling 120 kW identified by the heat flow at the bottom of the cascade.

2.2.3.6 Threshold problems

Threshold problems feature only one utility type – either hot or cold. They are important mostly because they often result in no utility–capital trade-off below a certain value of ΔT_{min}, since the minimum utility demand (hot or cold) becomes invariant; see Fig. 2.33.

Typical examples of threshold Heat Integration problems involve high-temperature fuel cells, which usually have large cooling demands but no heating demands – some examples are the flowsheets of Molten Carbonate Fuel Cells (Varbanov et al., 2006) and Solid Oxide Fuel Cells (Varbanov and Klemeš, 2008). An essential feature that distinguishes threshold problems is that, as ΔT_{min} is varied, demands for only one utility type (hot or cold) are identified over the variation range; in contrast, Pinched problems require both hot and cold utilities over this range. When synthesising HENs for threshold problems, one can distinguish between two subtypes (see Fig. 2.34):

1. Low-threshold ΔT_{min}: Problems of this type should be treated exactly as Pinch-type problems.
2. High-threshold ΔT_{min}: For these problems, it is first necessary to satisfy the required temperature for the no-utility end before proceeding with the remaining design.

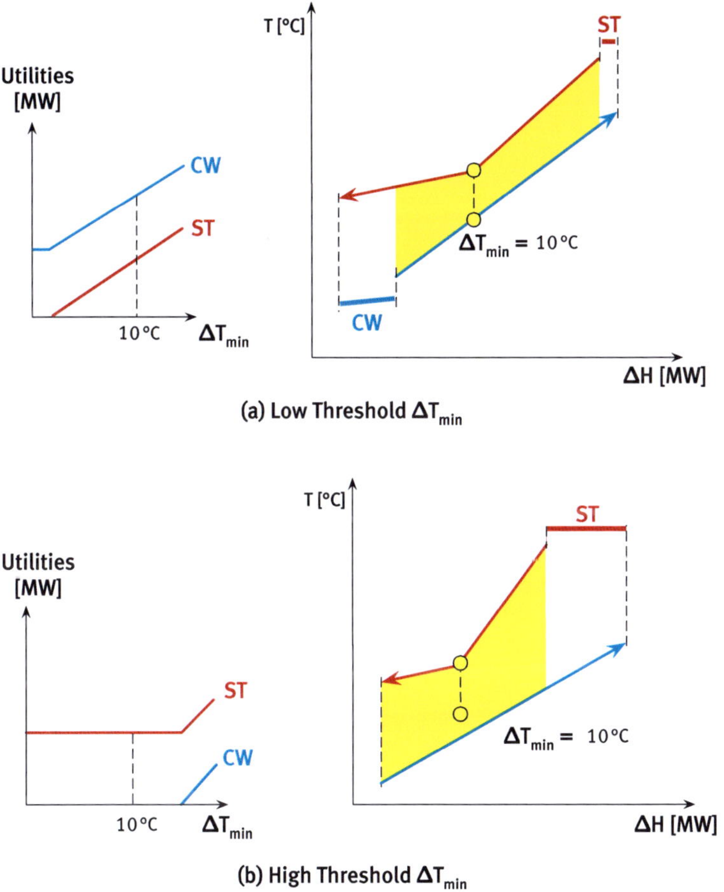

Fig. 2.33: Threshold problems.

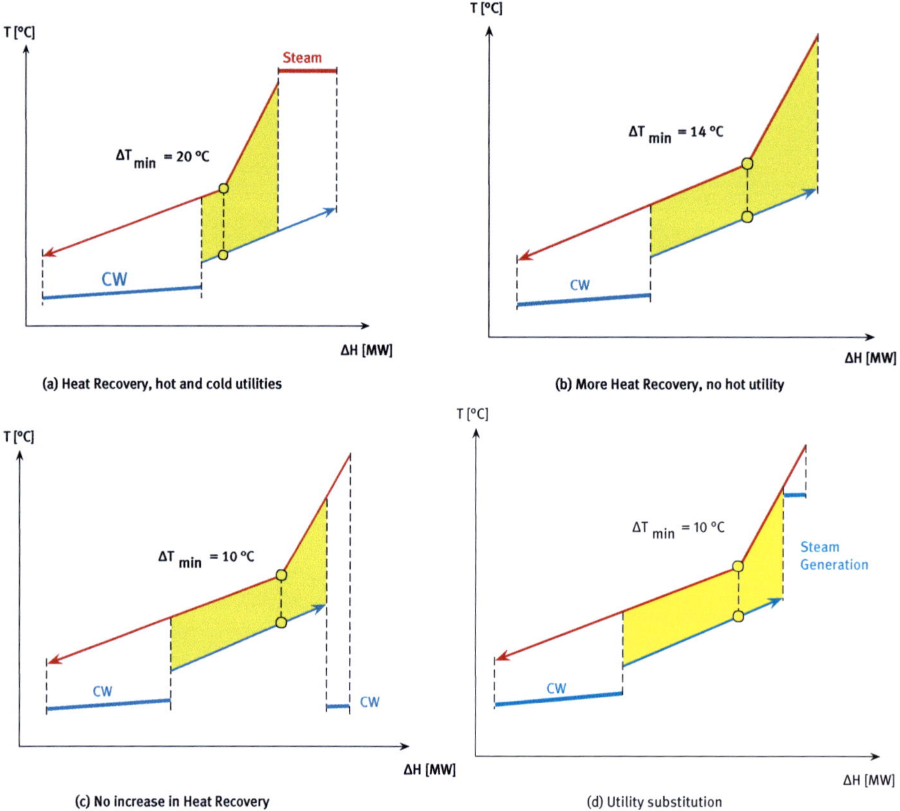

Fig. 2.34: Threshold HEN design cases.

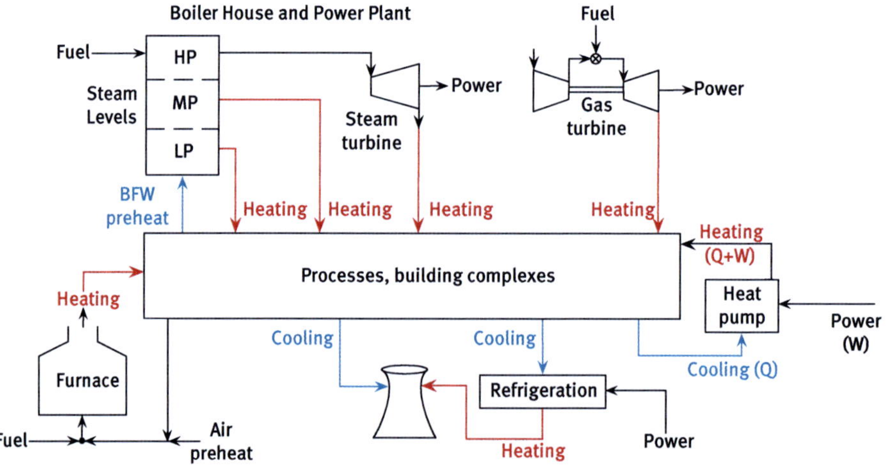

Fig. 2.35: Choices of hot and cold utilities (amended after CPI 2004 and 2005).

2.2.3.7 Multiple utilities targeting
Utility placement: Grand Composite Curve (GCC)

In many cases, more than one hot and one cold utility are available to provide the external heating and cooling requirements after energy recovery. It thus becomes necessary to find the cheapest and most effective combination of the available utilities (Fig. 2.35).

To assist with this choice and to enhance the information derived from the HCC and CCC, another graphical construction has been developed: the Grand Composite Curve (GCC) (Townsend and Linnhoff, 1983). The heat cascade and the PTA (Linnhoff and Flower, 1978) offer guidelines for the optimum placement of hot and cold utilities, and these allow determining the heat loads associated with each utility. To this point, the assumption has been that only one cold and one hot utility are available – albeit with sufficiently low and high temperatures to satisfy the cooling and/or heating demands of the process. However, most industrial sites feature multiple heating and cooling utilities at several different temperature levels (e.g., steam levels, refrigeration levels, hot oil circuit, furnace flue gas). Each utility has a different unit cost. Usually the higher temperature hot utilities and the lower temperature cold utilities cost more than the ones with temperatures closer to the ambient. This fact underscores the need to choose a mix resulting in the lowest utility cost. The general objective is to maximise the use of cheaper utilities and to minimise the use of more expensive utilities. For example, it is usually preferable to use low-pressure (LP) instead of high-pressure (HP) steam and to use cooling water instead of refrigeration. The Composite Curves plot (CC) (Fig. 2.18) provides a convenient view for evaluating the process driving force and the general Heat Recovery targets. However, the CC is not useful for identifying targets when multiple utility levels are available; the GCC is used for this task.

Construction of the Grand Composite Curve

The GCC is constructed using the Problem Heat Cascade (Fig. 2.31). The heat flows are plotted in the T-ΔH space, where the heat flow at each temperature boundary corresponds to the X coordinate and the temperature to the Y coordinate (Fig. 2.36).

The Grand Composite Curve can be directly related to the Shifted Composite Curves (SCCs), which are the result of shifting the CCs toward each other by $\Delta T_{min}/2$ so that the curves touch each other at the Pinch; see Fig. 2.37. At each temperature boundary, the heat flow in the Problem Heat Cascade and GCC corresponds to the horizontal distance between the SCCs.

The GCC has several fundamental properties that facilitate an understanding of the underlying Heat Recovery problem. The parts with positive slope (i.e., running uphill from left to right) indicate that cold streams dominate (Fig. 2.36 and Fig. 2.37). Similarly, the parts with negative slope indicate excess hot streams. The shaded areas, which signify opportunities for process-to-process Heat Recovery, are referred to as *Heat Recovery pockets*.

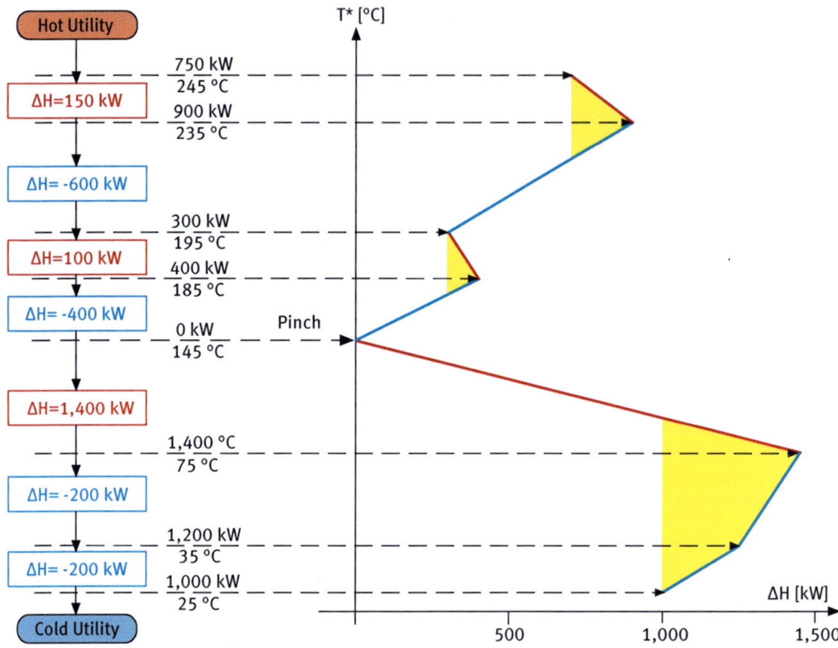

Fig. 2.36: Constructing the GCC for the streams in Table 2.3.

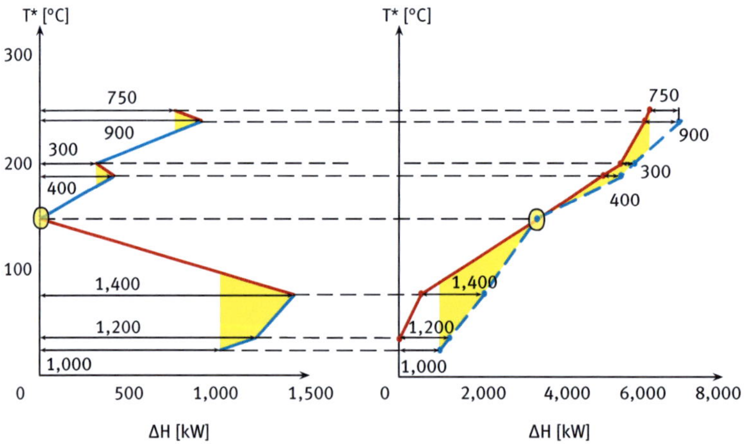

Fig. 2.37: Relation between the GCC (left) and the SCC (right) for the streams in Table 3.

Utility placement options

The GCC shows the hot and cold utility requirements of the process in terms of both enthalpy and temperature. This allows one to distinguish between utilities at different temperature levels. There are typically utilities at several different temperature levels available on a site. For example, it can be a supply of both high-pressure (HP) and medium-pressure (MP) steam. As indicated previously, it is desirable to maximise the use of cheaper utilities and to minimise the use of more expensive ones. Utilities at higher temperature and/or pressure are usually more expensive; see Fig. 2.38. Therefore, MP steam is used first: ranging from the y-axis until it touches the GCC, which maximises its usage. Only then is HP steam used.

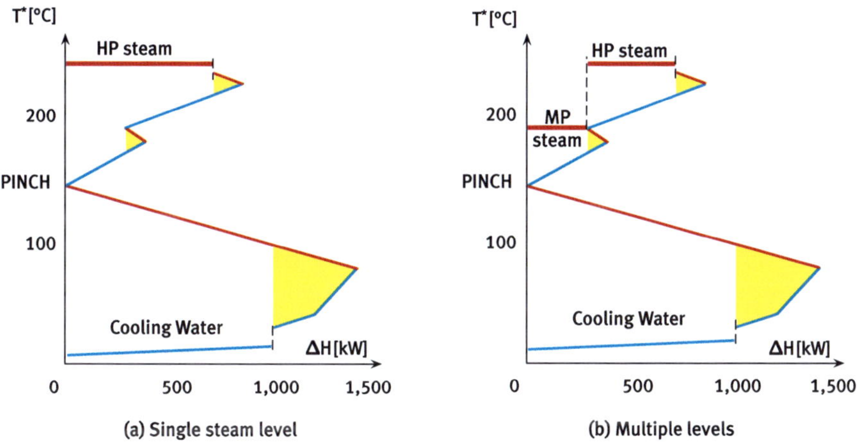

Fig. 2.38: Targeting for two steam levels using the GCC.

When a utility line or profile touches the GCC, a new Pinch Point is created, termed a "Utility Pinch" (Fig. 2.38). Each additional steam level creates another Utility Pinch and so increases the complexity of the utility system. Higher complexity has several negative consequences, including increased capital costs, greater potential for leaks, reduced safety, and more maintenance expenses. Limits are typically placed on the number of steam levels.

Higher temperature heating demands are satisfied by non-isothermal utilities. These include hot oil and hot flue gas, both of which maintain their physical phase (liquid and gaseous, respectively) across a wide range of temperatures. The operating costs associated with such utilities are largely dependent on furnace efficiency and on the intensity and efficiency of the pumping or fan blowing. When targeting the placement of a non-isothermal hot utility, its profile is represented by a straight line, which runs from the upper right to the lower left in the graph of Fig. 2.39. The line's starting point corresponds to the utility supply temperature and also to the rightmost point for the utility's heating duty. The utility use endpoint corresponds either to the zero of the ΔH axis – in which case all utility heating is covered by the current non-isothermal utility – or to the rightmost point on the ΔH axis for other, cheaper hot utilities.

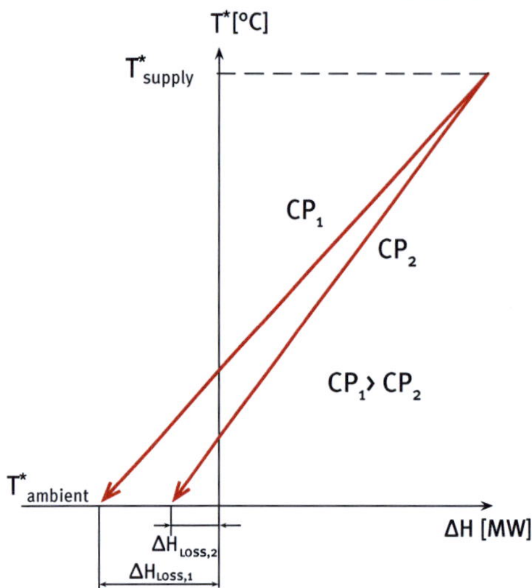

Fig. 2.39: Properties of non-isothermal hot utilities.

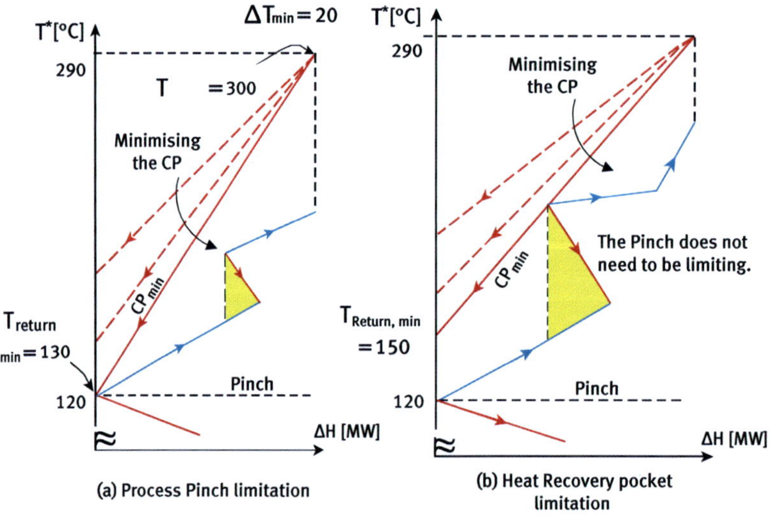

Fig. 2.40: Constraints for placing hot oil utilities.

As plotted in the figure, the non-isothermal utility's termination point corresponds to the ambient temperature. Thus, the distance from this point to the zero of the ΔH axis represents the thermal losses from using the utility. The heat capacity flow rate of the non-isothermal utility target is determined by making the utility line as steep as possible, thereby minimising its CP and the corresponding losses (Fig. 2.39). Its supply temperature is usually fixed at the maximum allowed by the furnace and the heat carrier composition; the remaining degree of freedom corresponds to the utility's exact CP. Smaller CP values result in steeper slopes and smaller losses. The placement for a non-isothermal utility (e.g. hot oil) may be constrained by two problem features: the process Pinch point and a "kink" in the GCC at the top end of a Heat Recovery pocket; see Fig. 2.40 for an example.

When fuel is burned in a furnace or a boiler, the resulting flue gas becomes available to heat up the corresponding cold-stream medium (for steam generation or direct process duty). Transferring heat to the process causes the flue gas temperature to drop as it moves from the furnace to the stack. The stack temperature has to be above a specified value: the *minimum allowed stack temperature*, which is determined by limitations due to corrosion. If flue gas is used directly for heating, then the Pinch point, if it is higher, may become more limiting than the minimum allowed stack temperature. If the analysed process features both high-temperature and moderate-temperature utility heating demands, then flue gas heating may not be appropriate for satisfying all those demands. If steam is cheaper, then combining it with flue gas reduces the latter's CP and the corresponding stack heat losses.

Another option for utility placement is to use part of the cooling demand of a process for generating steam. This is illustrated in Fig. 2.41, in which steam generation is placed below the Pinch.

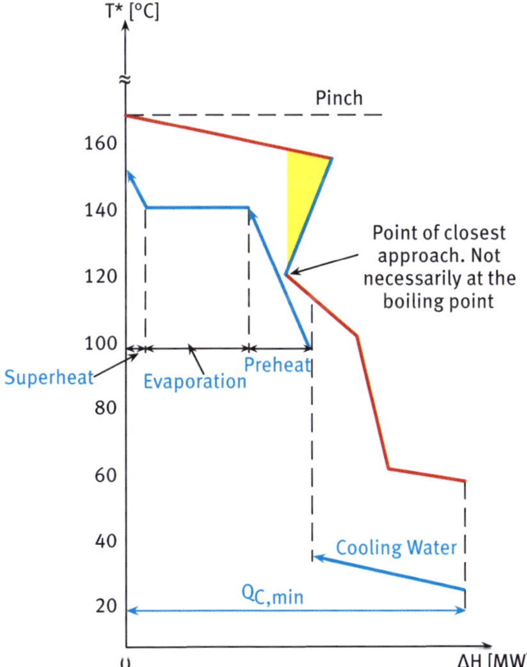

Fig. 2.41: Generating steam below the Pinch.

The GCC can reveal where utility substitution may improve energy efficiency; see Fig. 2.42. The main idea is to exploit Heat Recovery pockets that span two or more utility temperature levels. The technical feasibility of this approach is determined by both the temperature span and the heat duty within the pocket, which should be large enough (in terms of both heat duty and temperature span) to make utility substitution worthwhile when weighed against the required capital costs.

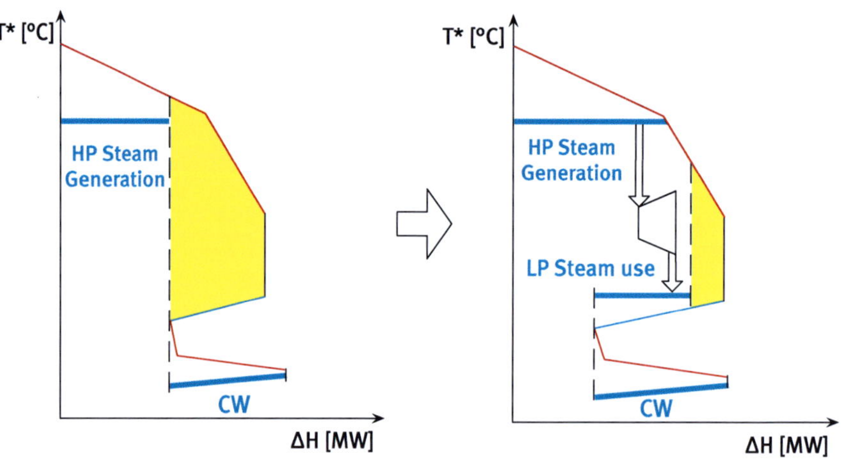

Fig. 2.42: Exploiting the pocket of the Grand Composite Curve for utility substitution.

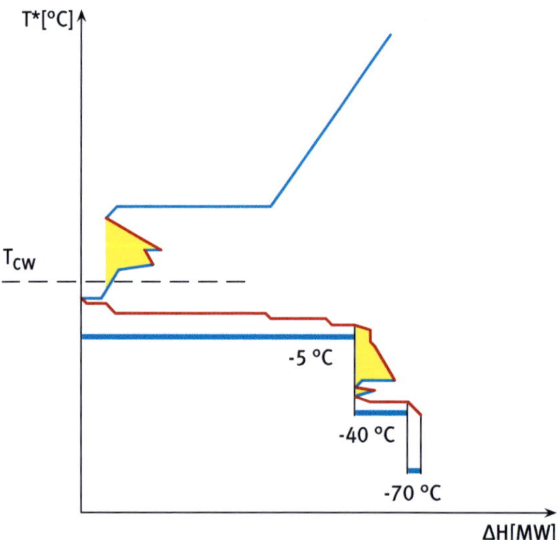

Fig. 2.43: Placing refrigeration levels for pure refrigerants.

Utility cooling below ambient temperatures may be required, a need that is usually met by refrigeration. Refrigerants absorb heat by evaporation, and pure refrigerants evaporate at a constant temperature. Therefore, refrigerants are represented – on the plot of T (or T*) versus ΔH – by horizontal bars, similarly to steam levels. On the GCC, refrigeration levels are placed similarly to steam levels; see Fig. 2.43.

When the level of a placed utility is between the temperatures of a Heat Recovery pocket, the Utility Pinch cannot be located by using the GCC. In this case, the Balanced Composite Curves (BCCs) are used. Fig. 2.44 shows how the data about the placed utilities can be transferred from the GCC to the BCCs, thus enabling correct location of the Utility Pinch associated with LP steam.

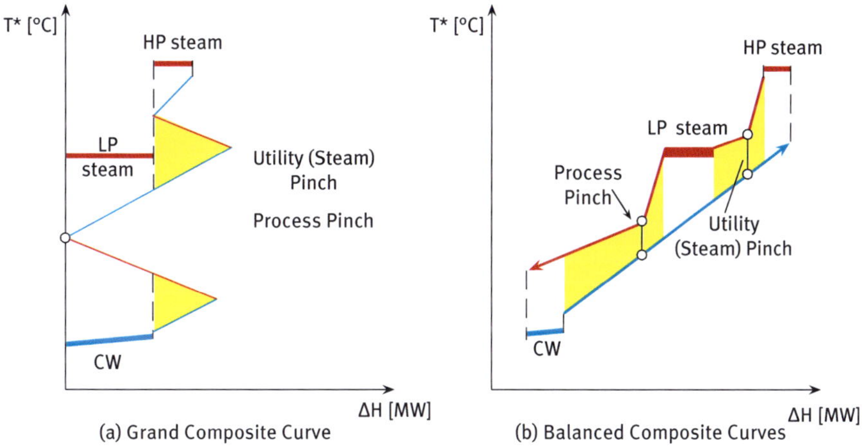

Fig. 2.44: Locating the LP steam Utility Pinch.

The BCCs create a combined view in which all heat sources and sinks (including utilities) are in balance and all Pinches are shown. Balanced Composite Curves are a useful additional tool for evaluating Heat Recovery, obtaining targets for specific utilities, and planning HEN design regions.

2.2.3.8 Investment and total cost targeting

In addition to estimating maximum Heat Recovery, it is also possible to estimate the required capital cost. The expressions for obtaining these estimates are derived from the relationship between heat transfer area and the efficiency of a single heat exchanger. Methods for targeting capital cost and total cost were initially developed by Townsend and Linnhoff (1984) and further elaborated by others – e.g., using simplified capital cost models (Linnhoff and Ahmad, 1990); using detailed capital cost models (Ahmad et al., 1990); using NLP transhipment models and account for forbidden matches –

(Colberg and Morari, 1990); integrated HEN targeting and synthesis within the context of the so-called "block decomposition" for HEN synthesis – (Zhu et al., 1995).

The HEN capital cost depends on the heat transfer area, the number of the heat exchangers, the number of shell-and-tube passes in each heat exchanger, construction materials, equipment type, and operating pressures. The heat transfer area is the most significant factor, and assuming one-shell-one-tube-pass exchangers it is possible to estimate the overall minimum required heat transfer area; this value helps establish the lower bound on the network's capital cost. Estimating the minimum heat transfer area is based on the concept of an enthalpy interval. As shown in Fig. 2.45, an *enthalpy interval* is a slice constrained by two vertical lines with fixed values on the ΔH axis. This interval is characterised by its enthalpy difference, the corresponding temperatures of the CCs at the interval boundaries, its process stream population, and the heat transfer coefficients of those streams.

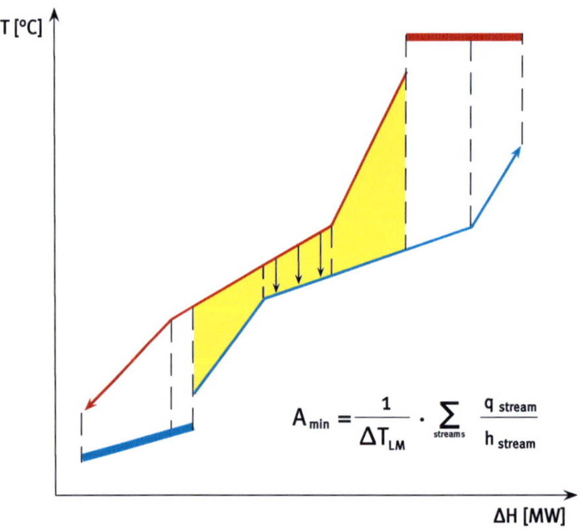

Fig. 2.45: Enthalpy intervals and area targeting.

The minimum heat transfer area target can be obtained by estimating it within each enthalpy interval of the BCC and then summing up the values over all intervals (Linnhoff and Ahmad, 1990):

$$A_{HEN,\,min} = \sum_{i=1}^{EI} \left[\frac{1}{\Delta T_{LM,i}} \cdot \sum_{s}^{NS} \frac{q_{s,i}}{h_{s,i}} \right] \qquad (2.12)$$

Here i denotes i-th enthalpy interval; s, the s-th stream; $\Delta T_{LM,i}$, the log-mean temperature difference in interval i (from the Composite Curve segments); q_s, the enthalpy change flow of the s-th stream part in the current enthalpy interval; and h_s, the heat transfer coefficient of s-th stream. The area targets can be supplemented by targets for number of shells (Ahmad and Smith, 1989) and for the number of heat exchanger units, thus providing a basis for estimating the HEN capital cost and the total cost. This approach is known as *supertargeting* (Ahmad et al., 1989).

With supertargeting it is also possible to optimise the value of ΔT_{min} prior to designing the HEN. Proposed improvements to the capital cost targeting procedure of Townsend and Linnhoff (1984) mainly involve: (1) obtaining more accurate surface area targets for HENs with non-uniform heat transfer coefficients – covered by references previously assessed (Colberg and Morari, 1990; Jegede and Polley, 1992; Zhu et al., 1995; Serna-Gonzalez et al., 2007); (2) accounting for construction materials, pressure rating, and different heat exchanger types (Hall et al., 1990); or (3) accounting for safety factors, including "prohibitive distance" (Santos and Zemp, 2000). Further information can be found in the paper by Taal et al. (2003), which summarises the common methods used to estimate the cost of Heat Exchange equipment and also provides sources of projections for energy prices.

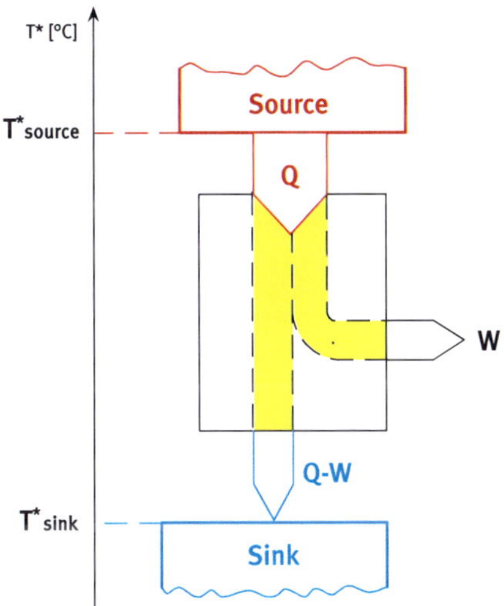

Fig. 2.46: Heat engine configuration.

2.2.3.9 Heat Integration of energy-intensive processes
Heat engines
Particularly important processes are the heat engines – steam and gas turbines as well as hybrids. They operate by drawing heat from a higher temperature source, converting part of it to mechanical power; then (after some energy loss) they reject the remaining heat at a lower temperature. See Fig. 2.46. The energy losses are usually neglected for targeting purposes.

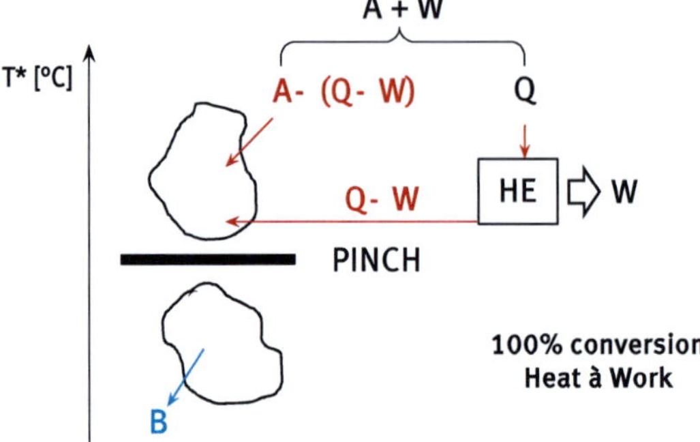

(a) Integration above the Pinch

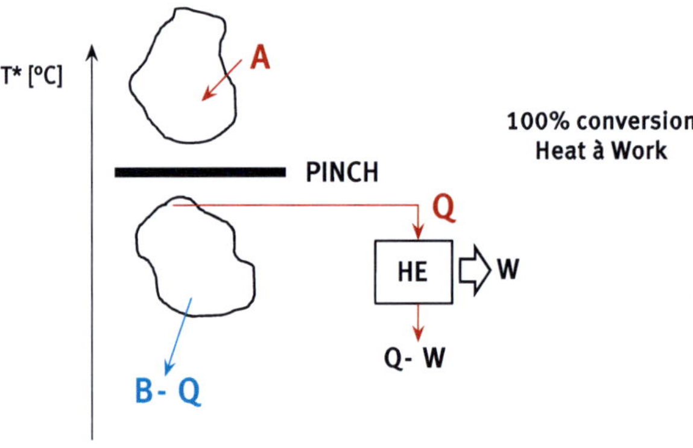

(b) Integration below the Pinch

Fig. 2.47: Appropriate placement of heat engines.

Integrating a heat engine across the Pinch, which is equivalent to a Cross-Pinch process to process heat transfer, results in a simultaneous increase of hot and cold utilities. This usually leads to excessive capital investment for the utility exchangers. If a heat engine is integrated across the Pinch, then the hot utility requirement is increased by Q and the cold by Q–W (according to the notation in Fig. 2.46). Heat engines should be integrated in one of two ways:

Above the Pinch (Fig. 2.47(a)): This increases the hot utility for the main process by W, but all the extra heat is converted into shaft work.

Below the Pinch (Fig. 2.47(b)): This offers a double benefit. It saves on a cold utility, and the process heat below the Pinch supplies Q to the heat engine (instead of rejecting it to a cooling utility).

From the perspective of the heat engines, their placement may differ. On the one hand, steam turbines may be placed either below or above the Pinch – since they draw and exhaust steam. Fig. 2.48 shows steam turbine integration above the Pinch, which has the benefit of cogenerating extra power. Gas turbines, on the other hand, which use fuel as input, are typically used only as a utility heat source for the processes and thus can be placed only above the Pinch.

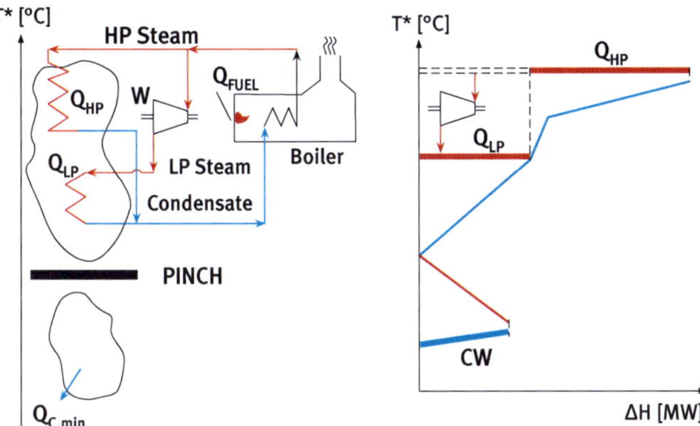

Fig. 2.48: Integrating a steam turbine above the Pinch.

Heat pumps

Heat pumps present another opportunity for improving the energy performance of an industrial process. Their operation is the reverse of heat engines. That is, heat pumps take heat from a lower temperature source, upgrade it by applying mechanical power, and then deliver the combined flow to a higher temperature heat sink (Fig. 2.49).

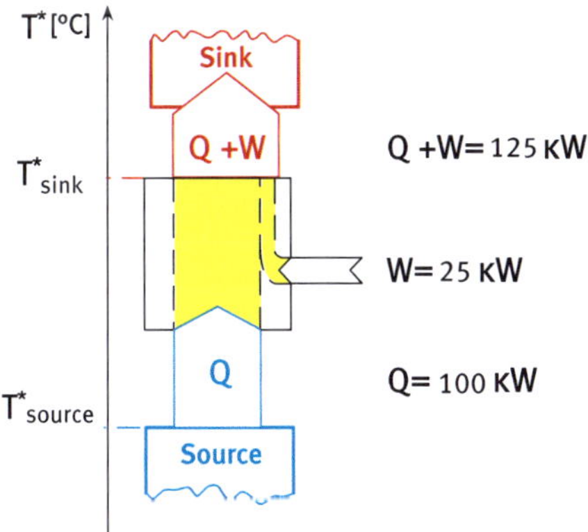

Fig. 2.49: Heat-pump configuration.

An important characteristic of heat pumps is their coefficient of performance (COP). This metric for device efficiency is defined as the ratio between the heat delivered to the heat sink and the consumed shaft work (mechanical power):

$$Q_{\text{sink}} = Q_{\text{source}} + W \tag{2.13}$$

$$COP = \frac{Q_{\text{sink}}}{W} = \frac{Q_{\text{source}} + W}{W}. \tag{2.14}$$

The COP is a nonlinear function of the temperature difference between the heat sink and the heat source (Laue, 2006); this difference is also referred to as *temperature lift*. Fig. 2.50(a) shows the appropriate integration of a heat pump across the Pinch, with the heat source located below the Pinch and the heat sink above it. The GCC facilitates sizing of the heat pump by evaluating the possible temperatures of the heat source and heat sink, and their loads; see Fig. 2.50(b). Integrating entirely above the Pinch results in direct conversion of mechanical power to heat. This is waste of primary resources, because most of the power is generated at the expense of twice to three times the amount of primary fuel energy. The second alternative – placing the heat pump entirely below the Pinch – results in the power flow consumed by the heat pump being added to the cooling utility demand below the Pinch.

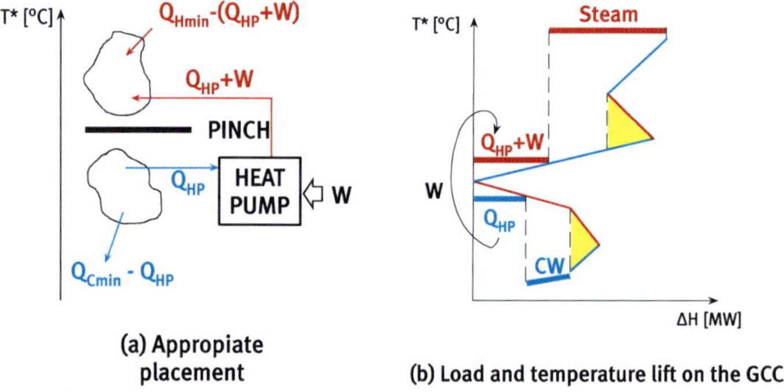

(a) Appropiate placement

(b) Load and temperature lift on the GCC

Fig. 2.50: Heat pump placement for a Heat Recovery problem.

The procedure for sizing heat pumps to be placed across a (process or utility) Pinch is illustrated in Fig. 2.51. First, temperatures are chosen for the heat source and the heat sink. Then the horizontal projections spanning from the temperature axis to the GCC provide the maximum values for the heat source and sink loads. Recall that the GCC shows shifted temperatures. Because true temperatures are used when calculating the actual heat-pump temperature lift, the GCC values must be modified by subtracting or adding $\Delta T_{\text{min}}/2$ (see Section 2.2.3.7). From the calculated ΔT_{pump} one can derive the COP, which is then used to calculate the necessary duties.

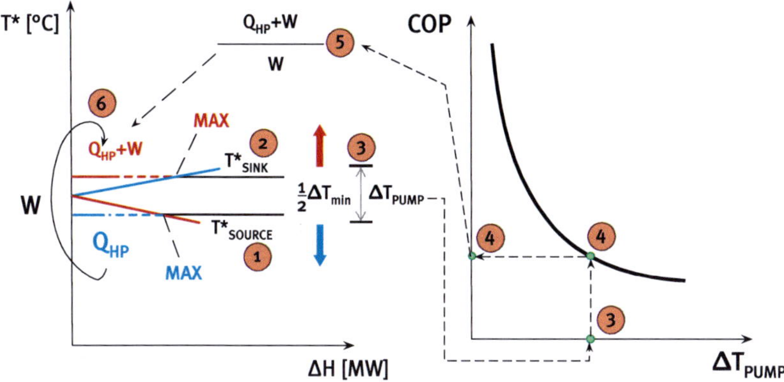

Fig. 2.51: Procedure for heat-pump sizing.

$\Delta T_{MIN} = 20°C$

ΔH [MW]	T*[°C]
21.90	440
29.40	410
23.82	131
18.00	118
1.80	115
0.00	94
4.30	91
11.50	79
15.00	30

Fig. 2.52: Heat-pump sizing example: Initial data.

As a concrete example, assume that the GCC in Fig. 2.52 reflects an industrial process with $\Delta T_{min} = 20°C$ and a heat pump is available, described as follows:

$$COP = 100.18 \cdot \Delta T_{pump}^{-0.874}. \tag{2.15}$$

Focusing on the Pinch "nose" (a sharp nose provides a better integration option, but watch also the scale of both axes) allows choosing a shifted temperature for the heat source, in this example: $T^*_{source} = 85\,°C$: see Fig. 2.53. Using this value allows one to extract a maximum of 6.9 MW from the GCC below the Pinch. Setting $T^*_{sink} = 100\,°C$ results in an upper bound of 2.634 MW for the sink load. Transforming to real temperatures and taking the difference yields a temperature lift of $\Delta T_{pump} = 35\,°C$. By Equation (2.15), the COP is thus equal to 4.4799. Given the upper bounds on the sink and source heat loads, the smaller one is chosen as a basis. Here the sink bound is smaller, so the sink is sized to its upper bound: $Q_{sink} = Q_{sink,max} = 2.634$ MW. From this, the required pump power consumption is computed to be 0.588 MW. As a result, the actual source load for the heat pump is 2.046 MW. Comparing this value with the upper bound of 6.9 MW, it is evident that the source heat availability is considerably underutilised.

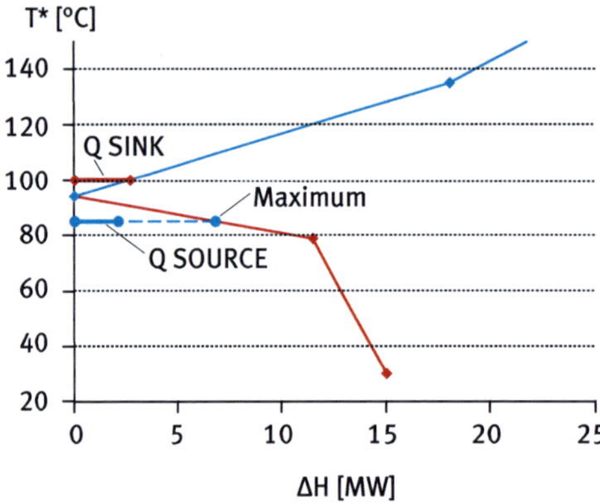

Fig. 2.53: Heat-pump sizing example: Attempt 1.

A different combination of source and sink temperatures is needed if the source availability is to be better utilised. As a second attempt, the sink temperature is increased from 100 to 110 °C. The maximum source heat remains 6.9 MW, but the maximum sink capacity increases from 2.634 to 7.024 MW. This results in the desired temperature lift of $\Delta T_{pump} = 45$ °C; the COP is now 3.5964, W = 2.657 MW, and the sink load $Q_{sink} = 9.557$ MW (Fig. 2.54).

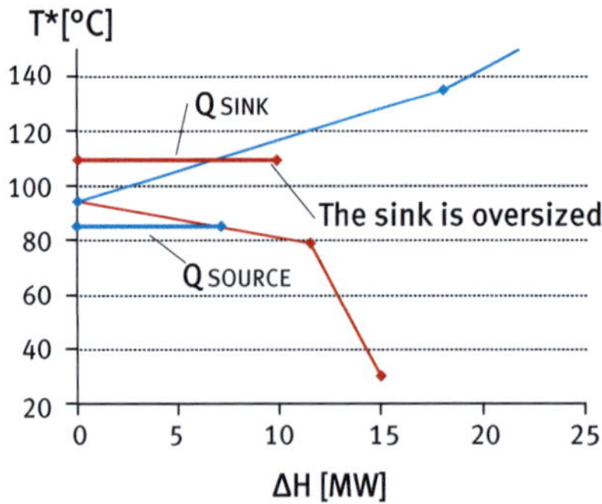

Fig. 2.54: Heat-pump sizing example: Attempt 2.

However, the heat sink is oversized in this new state (Fig. 2.54), so the heat source temperature must be shifted upward. An increase of 2.28 °C (see Fig. 2.55) yields the following values: ΔT_{pump} = 42.72 °C; a maximum source load (also taken as the actual source load) of 5.152 MW; a maximum sink load equal to the actual sink load of Q_{sink} = 7.016 MW; COP = 3.7637; and W = 1.864 MW. Better results can be obtained by optimising the two temperatures simultaneously while using overall utility cost as a criterion.

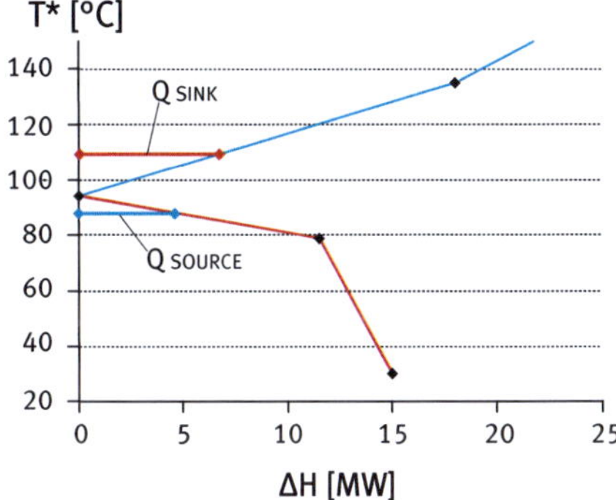

Fig. 2.55: Heat-pump sizing example: Attempt 3.

Once all utility levels are chosen, the heat pump can also be placed across a Utility Pinch (Fig. 2.56). A special case of integration involves the placement of refrigeration levels (Fig. 2.57). Refrigeration facilities are actually heat pumps whose primary value is that their cold end absorbs heat. Utilising their hot ends for process heating can save considerable amounts of hot utility, especially when relatively low-temperature heating is needed.

2.2.3.10 Distillation processes and other separators

The simplest distillation column includes one reboiler and one condenser, and these components account for most of the column's energy demands. For purposes of Heat Integration, the column is represented by a rectangle: the top side denotes the reboiler as a cold stream, and the bottom side denotes the condenser as a hot stream; see Fig. 2.58.

There are three options for integrating distillation columns: across the Pinch, as shown in Fig. 2.59(a); and entirely below or entirely above the Pinch, as shown in Fig. 2.59(b). Integrating across the Pinch only increases overall energy needs, so this option is not useful. The other two options result in net benefits by eliminating the need to use an external utility for supplying the distillation reboiler (below the Pinch) or the condenser (above the Pinch). The GCC is used to identify the appropriate

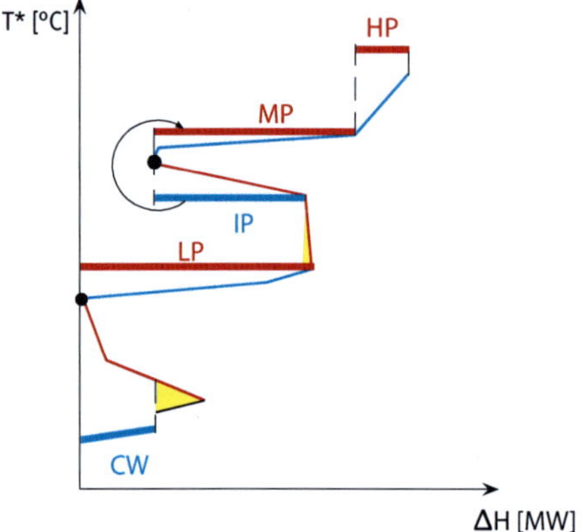

Fig. 2.56: Heat-pump placement across the Utility Pinch.

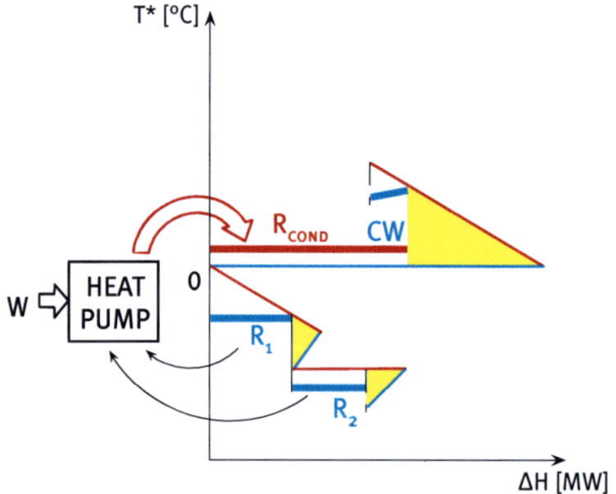

Fig. 2.57: Refrigeration systems.

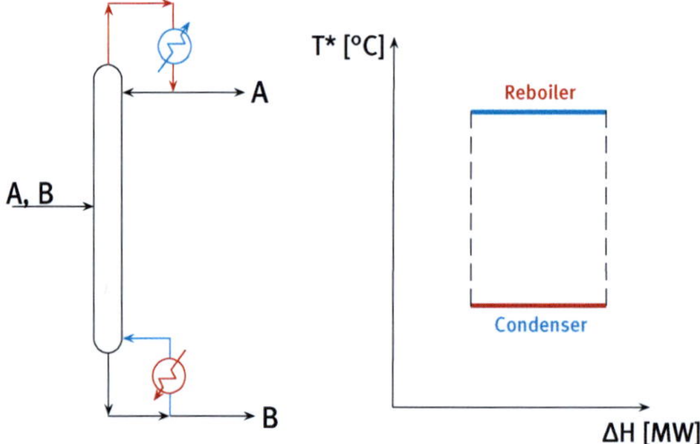

Fig. 2.58: Distillation column: T-H representation.

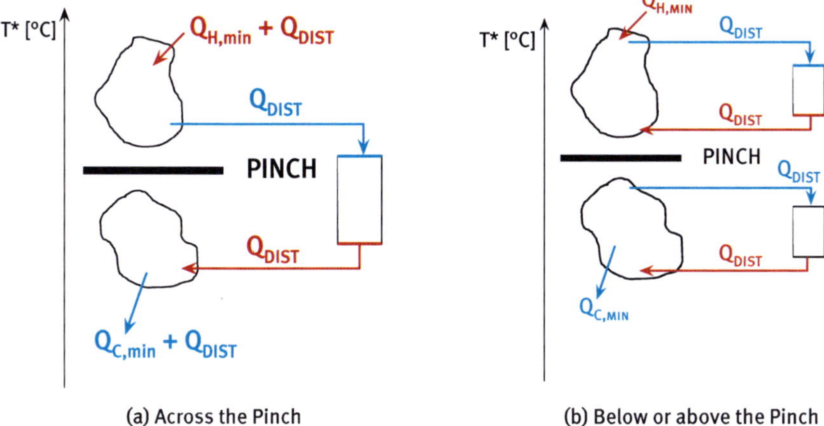

(a) Across the Pinch   (b) Below or above the Pinch

Fig. 2.59: Distillation column: Integration options.

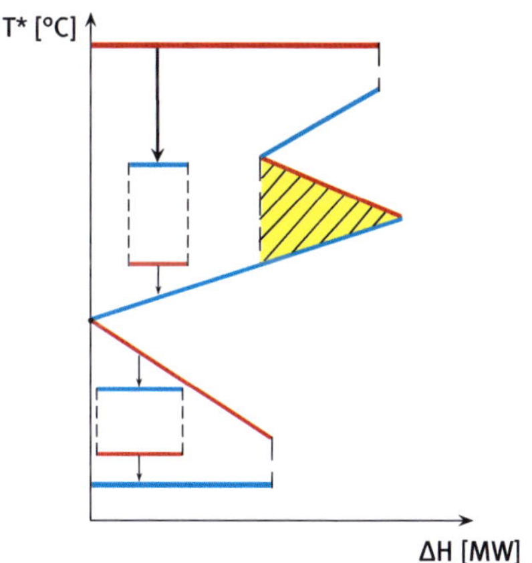

Fig. 2.60: Appropriate placement of distillation columns in terms of the GCC.

column integration options. Fig. 2.60 illustrates two examples of appropriately placed distillation columns.

With regard to the operating conditions of a column, if placed across the Pinch, several degrees of freedom can be utilised to facilitate appropriate placement. One option is to change the operating pressure, which would shift the column with respect to the temperature scale until it fits above or below the Pinch. The second degree of freedom is to vary the reflux ratio, which results in simultaneous changes of the column temperature span and the duties of the reboiler and condenser. Increasing the reflux ratio yields a smaller temperature span and larger duties, whereas reducing the ratio has the opposite effect. It is also possible to split the column into two parts, thereby introducing a "double effect" distillation arrangement. In this approach, one of the effects is placed below and the other above the Pinch, which prevents internal thermal integration of the column effects.

Additional options are available, such as inter-reboiling and inter-condensing. When using available degrees of freedom, one has to keep in mind that the energy–capital trade-offs of the column and the main process are combined and thus become more complicated than the individual trade-offs. Another important issue is controllability of the integrated designs: unnecessary complications should be avoided, and disturbance propagation paths should be discontinued. It is usually enough to integrate the reboiler or the condenser. If inappropriate column placement cannot be avoided, then it may be necessary to use condenser vapour recompression with a heat pump to heat up the reboiler.

Evaporators constitute another class of thermal separators. Because they also feature a reboiler and a condenser, their operation is similar to that of distillation columns. Therefore, the same integration principles can be applied to them. Absorber and stripper loops and dryers are similarly integrated.

Process modifications

The basic Pinch Analysis calculations assume that the core process layers in the Onion Diagram (Fig. 2.3) remain fixed. However, it is possible – and in some cases beneficial – to alter certain properties of the process. Properties that can be exploited as degrees of freedom include: (1) the pressure, temperature, or conversion rate in reactors; (2) the pressure, reflux ratio, or pump-around flow rate in distillation columns; and (3) the pressure of feed streams to evaporators and pressure inside evaporators.

Such modifications will also alter the heat capacity flow rates and temperatures of the related process streams for Heat Integration; this will cause further changes in the shapes of the CCs and the GCC, thereby modifying the utility targets. The CCs are a very valuable tool for suggesting beneficial process modifications (Linnhoff et al., 1982). Fig. 2.61 illustrates this application of the CCs in terms of the plus-minus principle. The main idea is to alter the CC's slope in the proper direction in order to reduce the amount of utilities needed. This can be achieved by changing CP (e.g., by mass flow variation). Such decreases in utility requirements can be brought about by (Smith, 2005) (1) increases in the total hot stream duty above the Pinch, (2) decreases in the total cold stream duty above the Pinch, (3) decreases in the total hot stream duty below the Pinch, and/or (4) increases in the total cold stream duty below the Pinch.

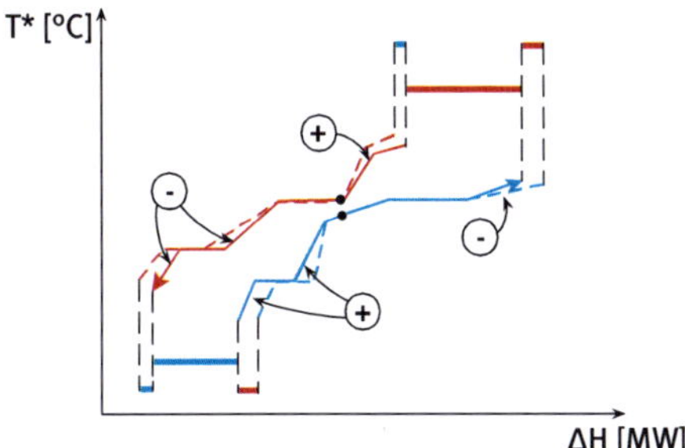

Fig. 2.61: The plus-minus principle.

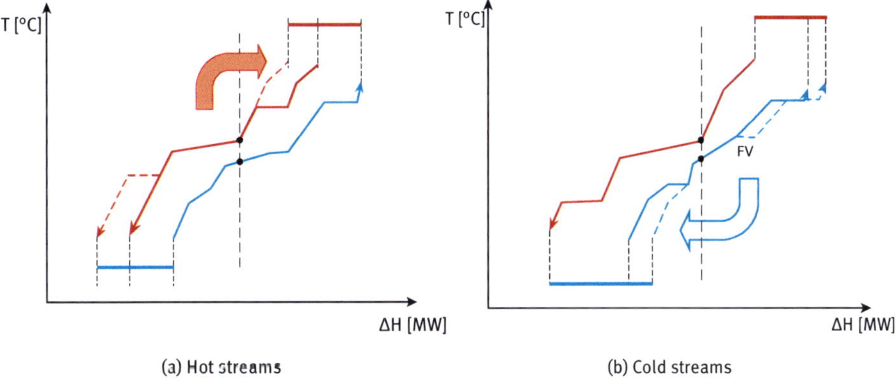

(a) Hot streams

(b) Cold streams

Fig. 2.62: Keep Hot Streams Hot (KHSH)/ Keep Cold Streams Cold (KCSC).

Another guide to modifying processes is the principle of *Keep Hot Streams Hot* (KHSH) and *Keep Cold Streams Cold* (KCSC). As illustrated in Fig. 2.62, increasing the temperature differences by process modification allows for more overlap of the curves and results in improved Heat Recovery. Energy targets improve if the heating and cooling demands can be shifted across the Pinch. The principle suggests (1) shifting hot streams from below to above the Pinch and/or (2) shifting cold streams from above to below the Pinch.

2.3 Summary

This chapter has provided an introduction to the most widespread methodology for evaluating Heat Recovery problems, combining as well as improving processes with the goal of minimising process utility demands. This is the methodology united under the title "Heat Integration". Applying Heat Integration allows identification of the thermodynamic limitations of a Heat Recovery problem in the form of the targets for Minimum Utility Heating, Minimum Utility Cooling and the Heat Recovery Pinch. These targets can be further used for completing related tasks:

– Estimation of the maximum possible performance of a Heat Exchanger Network (HEN) before designing it. This can be done by using the Composite Curves, the Grand Composite Curve or the Problem Table Algorithm.

– Synthesis and design of Heat Exchanger Networks (HENs). In this context the utility targets alone can be used as lower bounds on utility demands in mathematical-based optimisation formulations. The utility targets, combined with the Pinch location can be used for decomposing the Heat Recovery problem into several design areas, delimited by the process Pinch point as well the utilities at various levels defining Utility Pinches.

– Total Site Integration. This is an important extension of Heat Integration to complete industrial or combined sites, which include many industrial processes and/ or other energy users/generators, all of them linked by a common utility system. In this case the obtained process-level Heat Recovery targets can be used together to establish the site-wide utility targets and even evaluate the power cogeneration options.

References

Ahmad, S., Linnhoff, B. and Smith, R. (1989). Supertargeting: Different process structures for different economics, *Journal of Energy Resources Technology*, 111(3), 131–136.

Ahmad, S., Linnhoff, B. and Smith, R. (1990). Cost optimum heat exchanger networks: 2. Targets and design for detailed capital cost models, *Computers & Chemical Engineering*, 14(7), 751–767.

Ahmad, S. and Smith, R. (1989). Targets and design for minimum number of shells in heat exchanger networks, *Chemical Engineering Research and Design*, 67(5), 481–494.

Atkins, M.J., Walmsley, M.R.W. and Neale, J.R. (2010). The challenge of integrating non-continuous processes – milk powder plant case study, *Journal of Cleaner Production*, 18, 927–934.

Bochenek, R., Jezowski, J. and Jezowska, A. (1998). Retrofitting flexible heat exchanger networks – Optimization vs. sensitivity tables method. CHISA '98 / 1st Conference PRES '98 (Czech Society of Chemical Engineering, Prague), lecture F1.

Cerdá, J., Galli, M.R., Camussi, N. and Isla, M.A. (1990). Synthesis of flexible heat exchanger networks – I. Convex networks, *Computers & Chemical Engineering*, 14(2), 197–211.

Colberg, R.D. and Morari, M. (1990). Area and capital cost targets for heat exchanger network synthesis with constrained matches and unequal heat transfer coefficients, *Computers & Chemical Engineering*, 14(1), 1–22.

CPI (Centre for Process Integration) (2004). Heat integration and energy systems (MSc course, DPI UMIST), Manchester, UK.

Daichendt, M.M. and Grossmann, I.E. (1997). Integration of hierarchical decomposition and mathematical programming for the synthesis of process flowsheets, *Computers & Chemical Engineering*, 22(1–2), 147–175.

Foo D.C.Y., Chew, Y.H. and Lee, C.T. (2008). Minimum units targeting and network evolution for batch heat exchanger network, *Applied Thermal Engineering*, 28(16), 2089–2099.

Friedler, F. (2009). Process Integration, modelling and optimisation for energy saving and pollution reduction, *Chemical Engineering Transactions*, 18, 1–26.

Friedler, F. (2010). Process Integration, modelling and optimisation for energy saving and pollution reduction, *Applied Thermal Engineering*, 30(16), 2270–2280.

Furman, K.C. and Sahinidis, N.V. (2002). A critical review and annotated bibliography for heat exchanger network synthesis in the 20th century, *Industrial & Engineering Chemistry Research*, 41, 2335–2370.

Gogenko, A.L., Anipko O.B., Arsenyeva O.P. and Kapustenko P.O. (2007). Accounting for fouling in plate heat exchanger design, *Chemical Engineering Transactions*, 12, 207–212.

Hall, S.G., Ahmad, S. and Smith, R. (1990). Capital cost targets for heat exchanger networks comprising mixed materials of construction, pressure ratings and exchanger types, *Computers & Chemical Engineering*, 14(3), 319–335.

Hohmann, E.C. (1971). *Optimum networks for heat exchange*, PhD thesis, University of Southern California, Los Angeles, USA.

Jegede, F.O. and Polley, G.T. (1992). Capital cost targets for networks with non-uniform heat exchanger specifications, *Computers & Chemical Engineering*, 16(5), 477–495.

Klemeš, J., Friedler, F., Bulatov, I. and Varbanov, P. (2010). *Sustainability in the Process Industry – Integration and Optimization*, New York, USA: McGraw-Hill, 362 pp.

Laue, H.J. (2006). Heat pumps, in: Clauser, C., Strobl, T. and Zunic F. (eds.), *Renewable Energy*, Vol. 3C, Berlin, Germany: Springer.

Linnhoff, B. and Ahmad, S. (1990). Cost optimum heat exchanger networks – 1. Minimum energy and capital using simple models for capital cost, *Computers & Chemical Engineering*, 14(7), 729–750.

Linnhoff, B. and Flower, J.R. (1978). Synthesis of heat exchanger networks: I. Systematic generation of energy optimal networks, *AIChE Journal*, 24(4), 633–642.

Linnhoff, B. and Hindmarsh, E. (1983). The Pinch Design Method for Heat Exchanger Networks, *Chemical Engineering Science*, 38(5), 745–763.

Linnhoff, B., Townsend, D.W., Boland, D., Hewitt, G.F., Thomas, B.E.A., Guy, A.R. and Marsland, R.H. (1982). *A User Guide to Process Integration for the Efficient Use of Energy*, Rugby, UK: IChemE, latest edition 1994.

Masso, A.H. and Rudd, D.F. (1969). The synthesis of system designs II. Heuristic structuring, *AIChE Journal*, 15(1), 10–17.

Pan, M., Bulatov, I., Smith, R. and Kim J.-K. (2013a). Optimisation for the retrofit of large scale heat exchanger networks with different intensified heat transfer techniques, *Applied Thermal Engineering*, 53(2), 373–386.

Pan, M., Jamaliniya, S., Smith, R., Bulatov, I., Gough, M., Higley, T. and Droegemueller, P. (2013b). New insights to implement heat transfer intensification for shell and tube heat exchangers, *Energy*, DOI:10.1016/j.energy.2013.01.017.

Santos, L.C. and Zemp, R.J. (2000). Energy and capital targets for constrained heat exchanger networks, *Brazilian Journal of Chemical Engineering*, 17(4–7), 659–670.

Serna-González, M., Jiménez-Gutiérrez, A. and Ponce-Ortega, J.M. (2007). Targets for heat exchanger network synthesis with different heat transfer coefficients and non-uniform exchanger specifications, *Chemical Engineering Research and Design*, 85(10), 1447–1457.

Shah, R.K. and Sekulić, D.P. (2003). *Fundamentals of Heat Exchanger Design*. New York, USA: Wiley.

Shilling, R.L., Bell, K.J., Berhhagen, P.M., Flynn, T.M., Goldschmidt, V.M., Hrnjak, P.S., Standiford, F.C. and Timmerhaus, K.D. (2008). Heat-transfer equipment, in: Green, D.W. and Perry R.H. (eds.), *Perry's Chemical Engineers Handbook*, 8th ed, chap. 11, New York, USA: McGraw-Hill.

Smith, R. (2005). *Chemical Process Design and Integration*, Chichester, UK: John Wiley & Sons.

Taal, M., Bulatov, I., Klemeš, J. and Stehlik, P. (2003). Cost estimation and energy price forecast for economic evaluation of retrofit projects, *Applied Thermal Engineering*, 23, 1819–1835.

Ten Broeck, H. (1944). Economic selection of exchanger sizes, *Industrial & Engineering Chemistry Research*, 36(1), 64–67.

Tovazshnyansky, L.L., Kapustenko, P.O., Khavin, G.L. and Arsenyeva, O.P. (2004). *PHEs in Industry*, Kharkiv, Ukraine: NTU KhPI (in Russian).

Townsend, D.W. and Linnhoff, B. (1983). Heat and power networks in process design. Part II: Design procedure for equipment selection and process matching, *AIChE Journal*, 29(5), 748–771.

Townsend, D.W. and Linnhoff, B. (1984). Surface area targets for heat exchanger networks, IChemE 11th Annual Research Meeting (Bath, U.K., April).

Umeda, T., Harada, T. and Shiroko, K.A. (1979). Thermodynamic approach to the synthesis of heat integration systems in chemical processes, *Computers & Chemical Engineering*, 3(1–4), 273–282.

Varbanov, P. and Klemeš, J. (2008). Analysis and integration of fuel cell combined cycles for development of low-carbon energy technologies, *Energy*, 33(10), 1508–1517.

Varbanov, P., Klemeš, J., Shah, R.K. and Shihn, H. (2006). Power cycle integration and efficiency increase of molten carbonate fuel cell systems, *Journal of Fuel Cell Science and Technology*, 3(4), 375–383.

Zhu, X.X., O'Neill, B.K., Roach, J.R. and Wood, R.M. (1995). Area-targeting methods for the direct synthesis of heat exchanger networks with unequal film coefficients, *Computers & Chemical Engineering*, 19(2), 223–239.

Zhu, X.X., Zanfir, M. and Klemeš, J. (2000). Heat transfer enhancement for heat exchanger network retrofit, *Heat Transfer Engineering*, 21(2), 7–18.

3 Synthesis of Heat Exchanger Networks

Having established how much Heat Recovery can be achieved at most and having realised the thermodynamic limitations posed by the process streams, this information can be used to synthesise Heat Recovery systems using heat exchangers – Heat Exchanger Networks (HEN). This procedure is frequently referred to as Heat Exchanger Network Synthesis (HENS). There are plenty of methods for performing this task – some heuristic, others based on mathematical programming, as well as evolutionary ones using genetic algorithms and Simulated Annealing.

This chapter presents a systematic procedure for synthesising HENs using simple reasoning and calculations and thus allowing experienced engineers as well as beginners to apply it and benefit from the insights offered. This is the Pinch Design Method. Other synthesis methods are also considered, as well as the fundamentals of HEN retrofit.

3.1 Introduction

In Chapter 2 "Setting Targets and Heat Integration" Pinch Analysis was introduced. This is a powerful method based on sound thermodynamic concepts applied to the overall Heat Recovery problem for a process. In all cases the targets obtained by Pinch Analysis can be used as lower bounds on utility demand in mathematical programming procedures. However, the real power of the method is revealed when it is realised that the Pinch divides the problem into a net heat sink above the Pinch and a net heat source below the Pinch. This provides a means for problem decomposition – i.e. replacing one large and complex problem by an equivalent combination of two or more problems that are an order of magnitude simpler. A second benefit of the Pinch division is that the thermodynamic bottleneck identified by the Pinch, combined with a requirement for maximum Heat Recovery, also provides a set of rules for Heat Exchange match placement at the Pinch and further evolution of the design. This is the fundamental thinking behind the Pinch Design Method.

This chapter presents the Pinch Design Method in detail and then discusses other HEN synthesis methods as well as basic features of HEN retrofit. Working sessions are also provided to help improving and consolidating the understanding of the material.

3.2 HEN synthesis

Most industrial-scale methods synthesise Heat Recovery networks under the assumption of a steady state.

3.2.1 The Pinch Design Method

The traditional Pinch Design Method (Linnhoff and Hindmarsh, 1983) has become popular owing to its simplicity and efficient management of complexity. The method has evolved into a complete suite of tools for Heat Recovery and design techniques for energy efficiency, including guidelines for changing and integrating a number of energy-intensive processes.

3.2.1.1 HEN representation

The representation of Heat Exchanger Networks by a general process flowsheet, as in Fig. 3.1, is not convenient. The reason is that this representation makes it difficult to answer a number of important questions: "Where is the Pinch?" "What is the degree of Heat Recovery?" "How much cooling and heating from utilities is needed?"

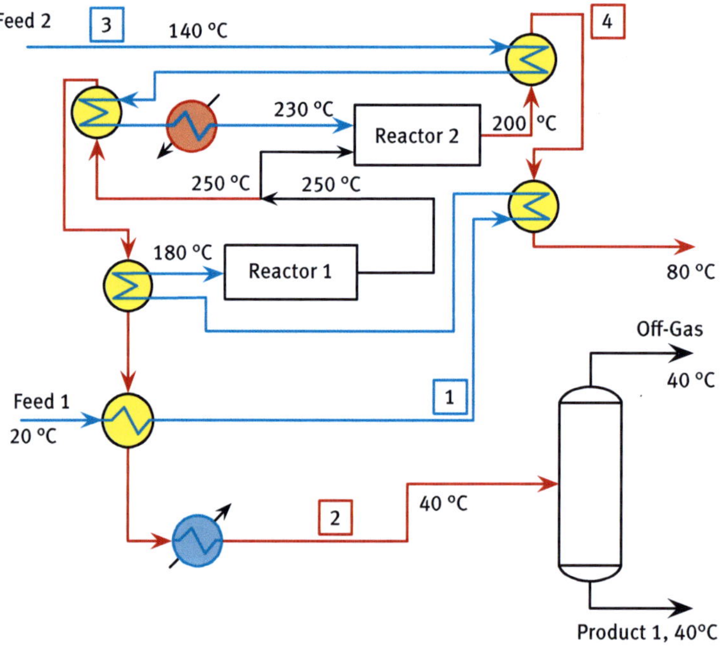

Fig. 3.1: Using a general process flowsheet to represent a HEN.

The example given in Fig. 3.1 results in the streams data set from Table 3.1. The so-called conventional HEN flowsheet (Fig. 3.2) offers a small improvement. It shows only heat transfer operations and is based on a simple convention: cold streams are depicted horizontally and hot streams vertically. Although the Pinch location can be marked for simple cases, it is still difficult to see. This representation makes it difficult to express the proper sequencing of heat exchangers and to represent clearly the network temperatures. Furthermore, changing the positions of some matches often results in complicated path representations.

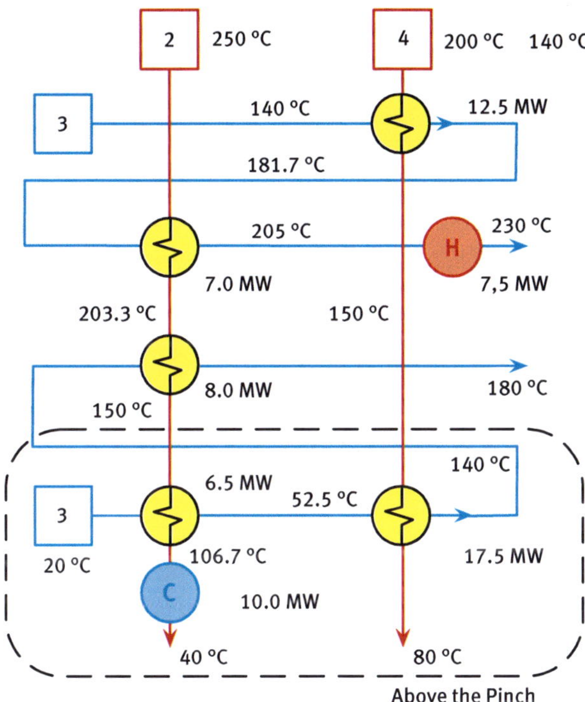

Fig. 3.2: Conventional HEN flowsheet.

Tab. 3.1: PTA example: Process streams data ($\Delta T_{min} = 10\,°C$).

No.	Type	T_S [°C]	T_T [°C]	CP [kW/°C]
1	Cold	20	180	20
2	Hot	250	40	15
3	Cold	140	230	30
4	Hot	200	80	25

The Grid Diagram, as shown in Fig. 3.3, provides a convenient and efficient representation of HENs by eliminating the problems just described. The Grid Diagram has several advantages: the representation of streams and heat exchangers is clearer, it is more convenient for representing temperatures, and the Pinch location (and its implications) are clearly visible; see Fig. 3.4. Once again, only heat transfer operations are shown. Temperature increases from left to right in the grid, which is intuitive, and (re)sequencing heat exchangers is straightforward.

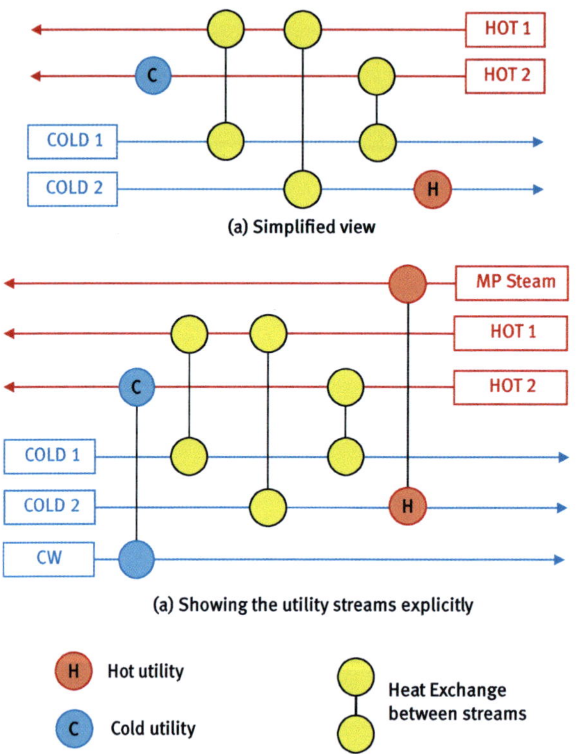

Fig. 3.3: The Grid Diagram for HENs.

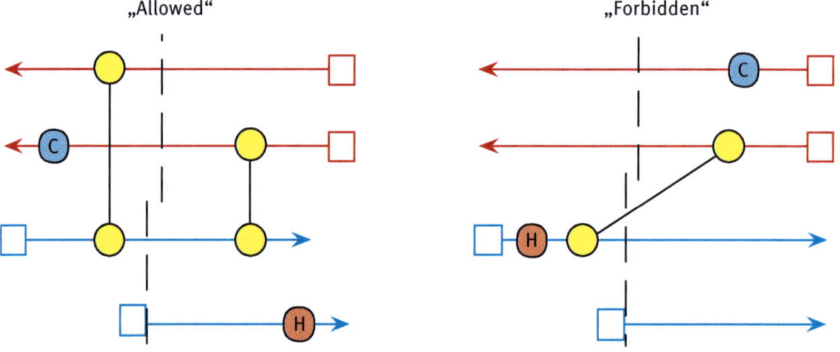

Fig. 3.4: The Grid Diagram and implications of the Pinch.

3.2.1.2 The design procedure

The procedure for designing a HEN follows several simple steps:
1. Specification of the Heat Recovery problem
2. Identification of the Heat Recovery targets and the Heat Recovery Pinch (Chapter 2 "Setting Targets and Heat Integration")
3. Synthesis
4. Evolution of the HEN topology

The first two steps are discussed in Chapter 2 "Setting Targets and Heat Integration". The synthesis step begins by dividing the problem at the Pinch and then positioning the process streams as shown in Fig. 3.5.

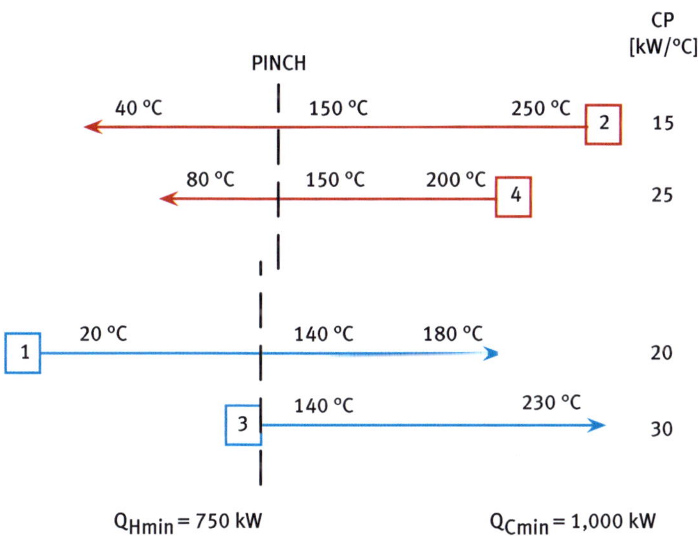

Fig. 3.5: Dividing at the Pinch for the streams from Table 3.1.

The Pinch design principle suggests starting the network design from the Pinch (the most restricted part of the design owing to temperature differences approaching ΔT_{min}) and then to place heat exchanger matches while moving away from the Pinch (Fig. 3.6). When placing matches, a few rules must be followed in order to obtain a network that minimises utility use: (1) no exchanger may have a temperature difference smaller than ΔT_{min}; (2) no process-to-process heat transfer may occur across the Pinch; and (3) no inappropriate use of utilities is allowed.

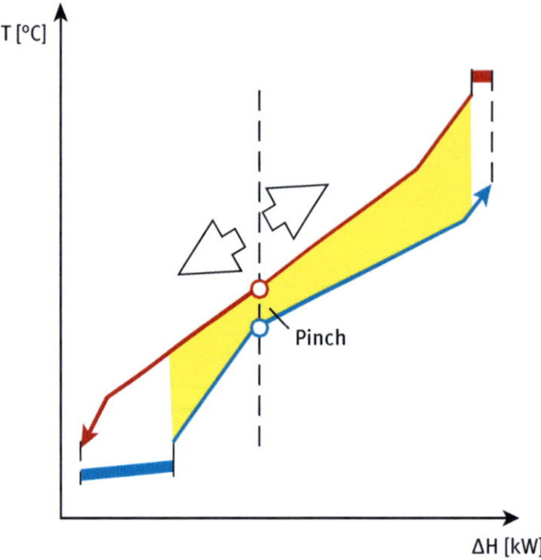

Fig. 3.6: The Pinch design principle.

At the Pinch, the driving force restrictions entail that certain matches must be made if the design is to achieve the minimum utility usage without violating the ΔT_{min} constraint; these are referred to as essential matches. Above the Pinch, the hot streams may be cooled only by transferring heat to cold process streams, not to utility cooling. Therefore, all hot streams above the Pinch have to be matched up with cold streams. This means that all hot streams entering the Pinch must be given priority when matches are made above the Pinch. Conversely, cold streams entering the Pinch are given priority when matches are made below the Pinch.

Now recall the example from Chapter 2 "Setting Targets and Heat Integration". It is repeated in Table 3.1. Fig. 3.5 shows the empty Grid Diagram, which is divided at the Pinch. The hot part (above the Pinch) requires essential matches for streams 2 and 4, since they are entering the Pinch.

Consider Stream 4. One possibility is to match it against Stream 1, as shown in Fig. 3.7. Stream 4 is the hot stream, and its CP is greater than the CP for cold Stream 1. As shown in the figure, at the Pinch the temperature distance between the two streams is exactly equal to ΔT_{min}. Moving away from the Pinch results in temperature convergence because the slope of the hot stream line is less steep owing to its larger CP. Since ΔT_{min} is the lower bound on network temperature differences, the proposed heat exchanger match is infeasible and thus is rejected.

Another possibility for handling the cooling demand of Stream 4 is to implement a match with Stream 3, as shown in Fig. 3.8. The CP of Stream 3 is larger than the CP of Stream 4, resulting in divergent temperature profiles in the direction away from the Pinch. The rule above the Pinch may be expressed as follows:

$$CP_{\text{hot stream}} \leq CP_{\text{cold stream}}. \tag{3.1}$$

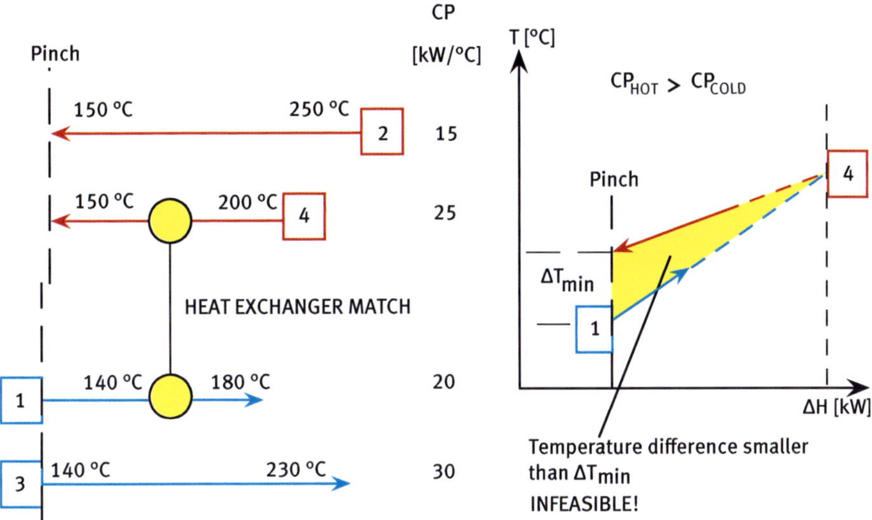

Fig. 3.7: An infeasible heat exchanger match above the Pinch.

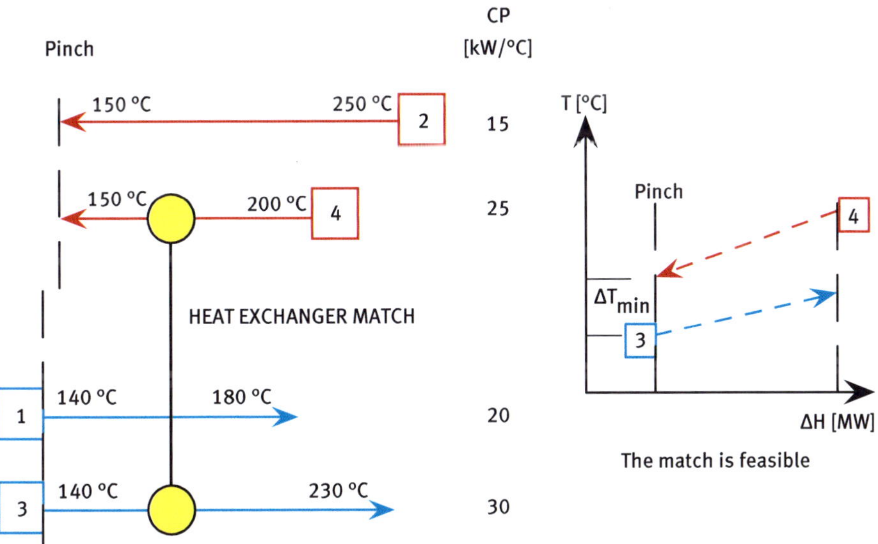

Fig. 3.8: A feasible heat exchanger match above the Pinch.

The picture below the Pinch is symmetric. This part of the design is a net heat source, which means that the heating requirements of the cold streams must be satisfied by matching up with hot streams. The same type of reasoning as before yields the following requirement: the CP value of a cold stream must not be greater than the CP value of a hot stream if a feasible essential match is to result. Generalising Equation (3.1) shows that the CP of the stream entering the Pinch has to be less than or equal to the CP of the stream leaving the Pinch:

$$CP_{\text{entering pinch}} \leq CP_{\text{leaving pinch}}. \qquad (3.2)$$

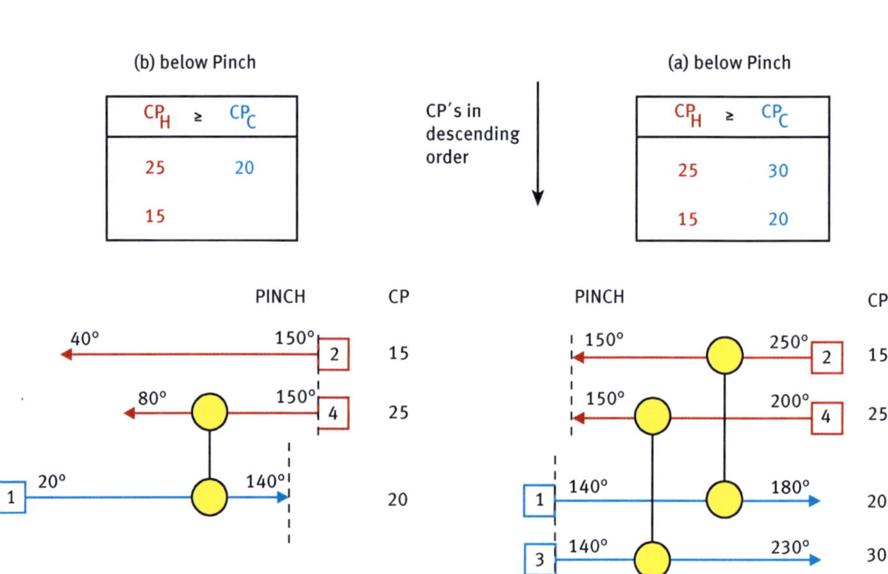

Fig. 3.9: CP tables.

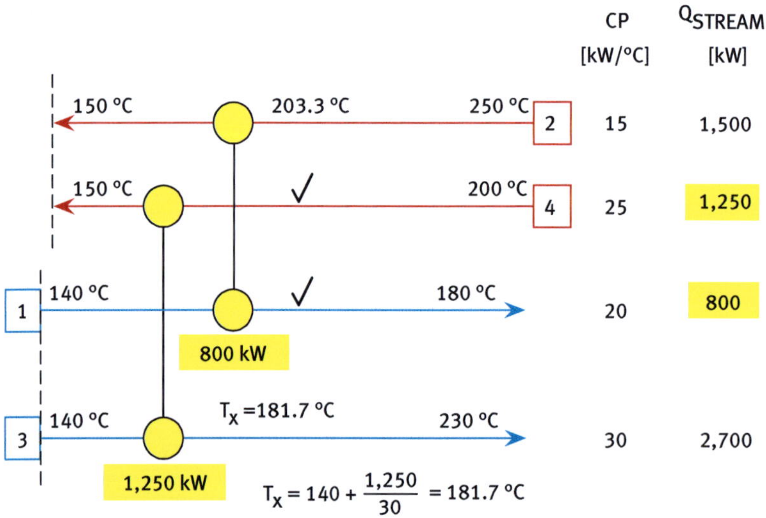

Fig. 3.10: The tick-off heuristic.

The Pinch Design Method incorporates a special tool for handling this stage: CP tables – Fig. 3.9 (Linnhoff and Hindmarsh, 1983). Here the streams are represented by their CP values, which are sorted in descending order. This facilitates the identification of promising combinations of streams as candidate essential matches.

Sizing the matches follows the tick-off heuristic, which stipulates that one of the streams in a Heat Exchange match should be completely satisfied and then "ticked off" from the design list; see Fig. 3.10.

3.2.1.3 Completing the design

The HEN design above the Pinch is illustrated in Fig. 3.11. The design below the Pinch follows the same basic rules, with the small difference that here it is the cold streams that define the essential matches.

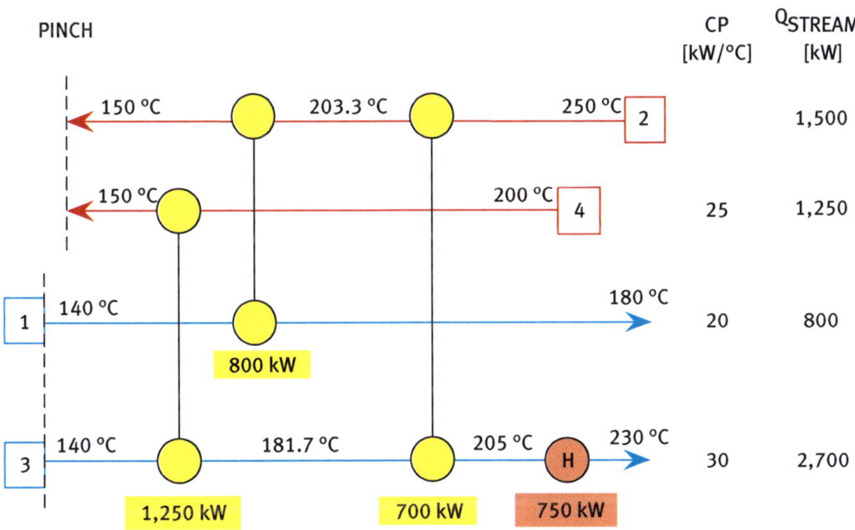

Fig. 3.11: Completing the HEN design above the Pinch.

Fig. 3.12 details the design below the Pinch for the considered example. First the match between Streams 4 and 1 is placed and sized to the duty required by Stream 4. The other match, between Streams 2 and 1, formally violates the Pinch rule for placing essential matches. However, since Stream 1 already has another match at the Pinch, the current match (between 4 and 1) cannot be strictly termed essential. Up to the required duty of 650 kW, this match does not violate the ΔT_{min} constraint, which is the relevant one. The completed HEN topology is shown in Fig. 3.13.

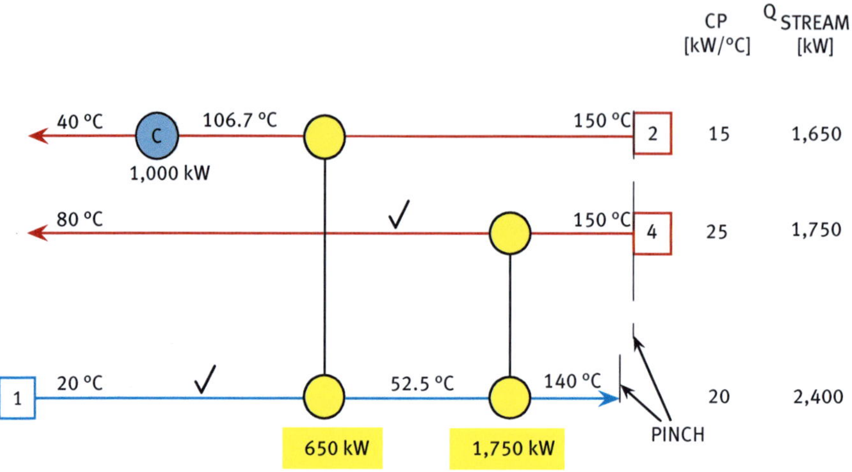

Fig. 3.12: The HEN design below the Pinch.

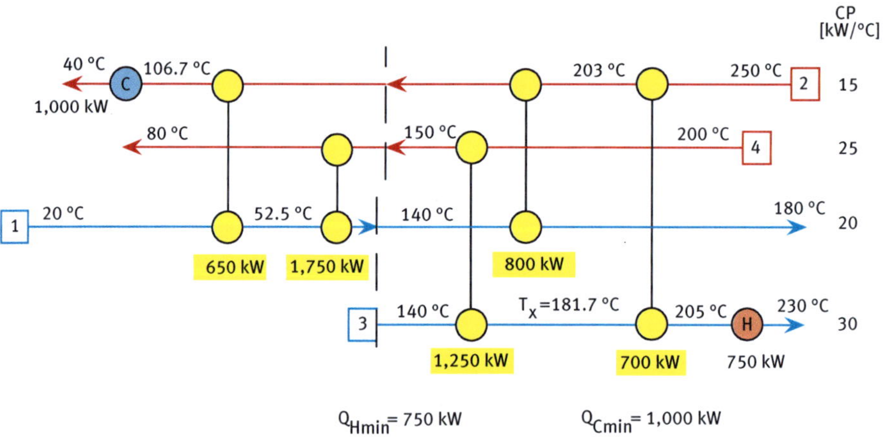

Fig. 3.13: The completed HEN design.

3.2.1.4 Stream splitting

The basic design rules described previously are not always sufficient. In some cases it is necessary to split the streams so that Heat Exchange matches can be appropriately placed. Splitting may be required in the following situations:

- Above the Pinch when the number of hot streams is greater than the number of cold streams ($N_H > N_C$); see Fig. 3.14.
- Below the Pinch when the number of cold streams is greater than the number of hot streams ($N_C > N_H$); see Fig. 3.15.
- When the CP values do not suggest any feasible essential match; see Fig. 3.16.

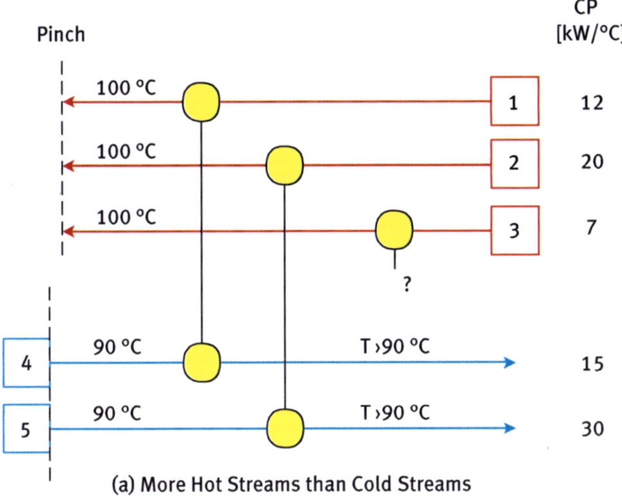

(a) More Hot Streams than Cold Streams

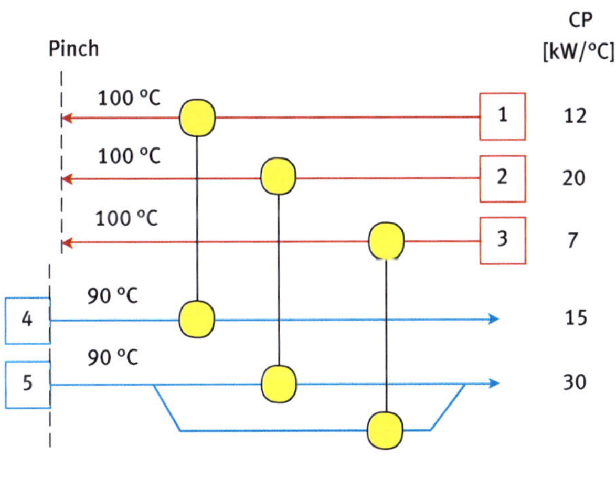

(b) Split a cold stream

Fig. 3.14: Splitting above the Pinch for $N_H > N_C$.

The loads of the matches involving stream branches are again determined using the tick-off heuristic. Because each stream splitter presents an additional degree of freedom, it is necessary to decide how to divide the overall stream CP between the branches. One possibility is illustrated in Fig. 3.17. The suggested split ratio there is 4 : 3. This approach would completely satisfy the heating needs of the cold stream branches.

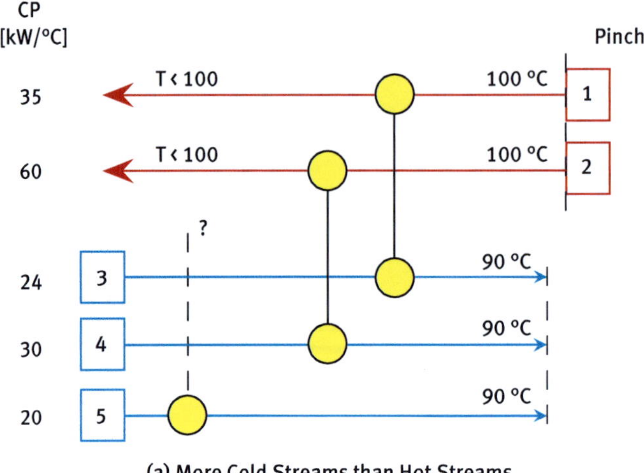

(a) More Cold Streams than Hot Streams

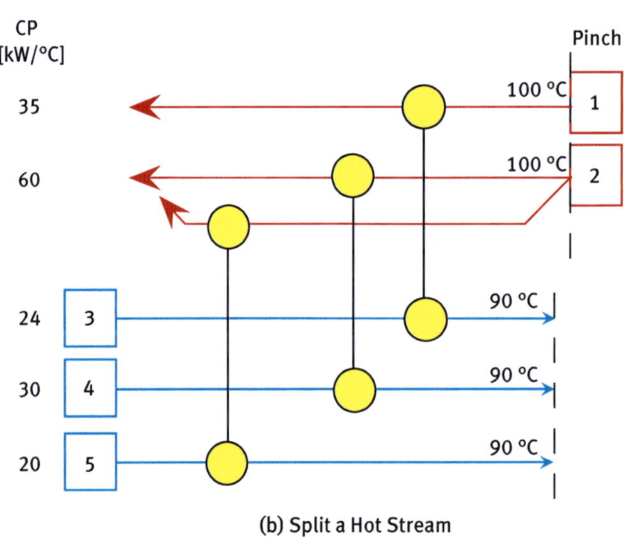

(b) Split a Hot Stream

Fig. 3.15: Splitting below the Pinch for $N_C > N_H$.

The arrangement in Fig. 3.17 is in a way trivial and can be improved on. Unless the stream combinations impose some severe constraints, there is an infinite number of possible split ratios. This fact can be exploited to optimise the network. In many cases it is possible to save an extra heat exchanger unit by ticking off two streams with a single match, as illustrated in Fig. 3.18, which also satisfies hot Stream 2 in addition to both branches of cold Stream 3, thereby eliminating the second cooler.

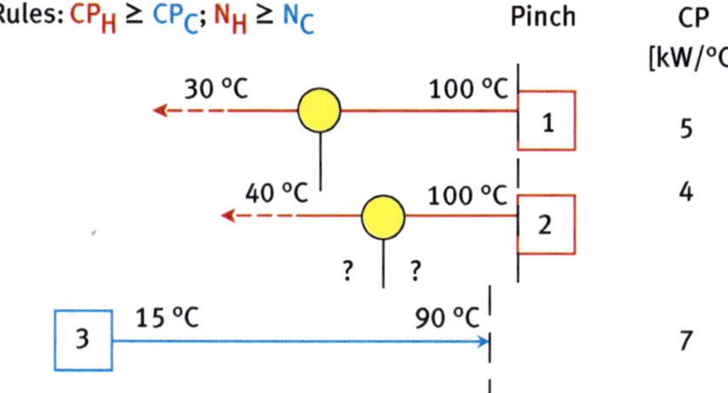

(a) CP rules are not satisfied

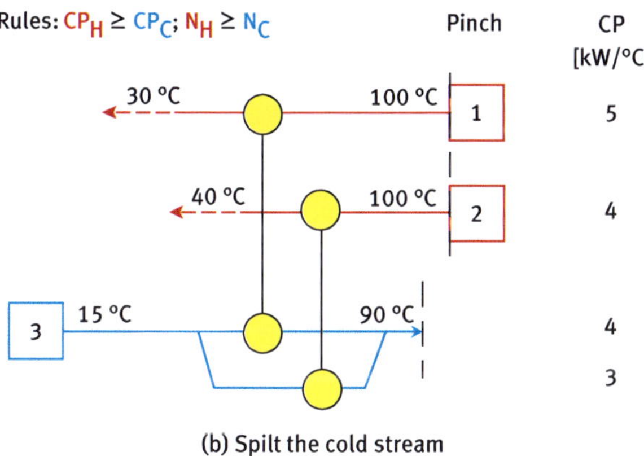

(b) Spilt the cold stream

Fig. 3.16: Splitting to enable CP values for essential matches.

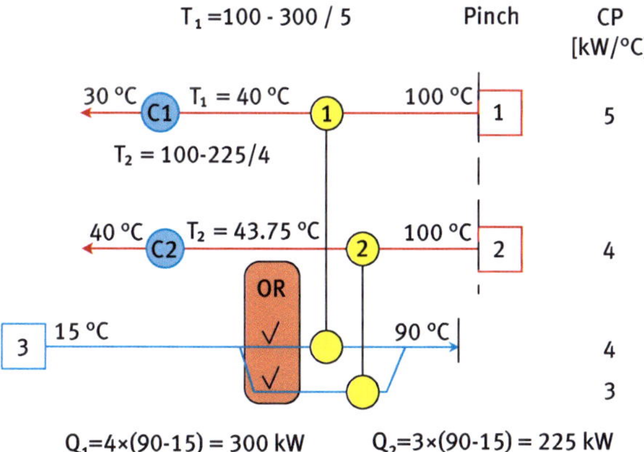

Fig. 3.17: Splitting and (trivial) tick-off.

The complete algorithm for splitting streams above the Pinch is given in Fig. 3.19. The procedure for splitting below the Pinch is symmetrical, with the cold and hot streams switching their roles.

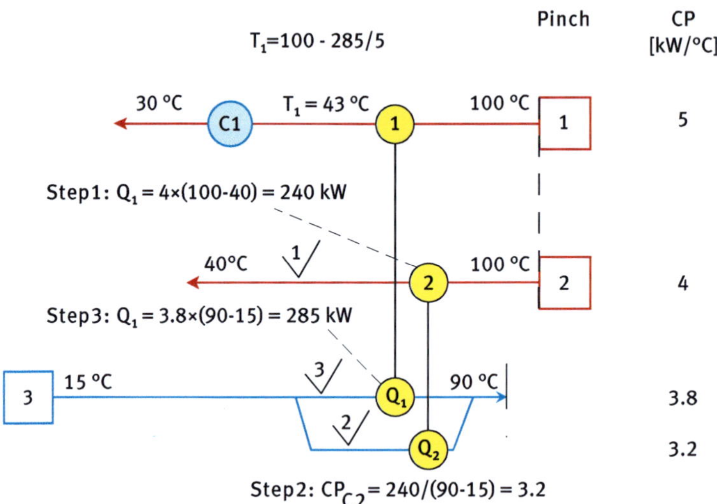

Fig. 3.18: Splitting and advanced tick-off.

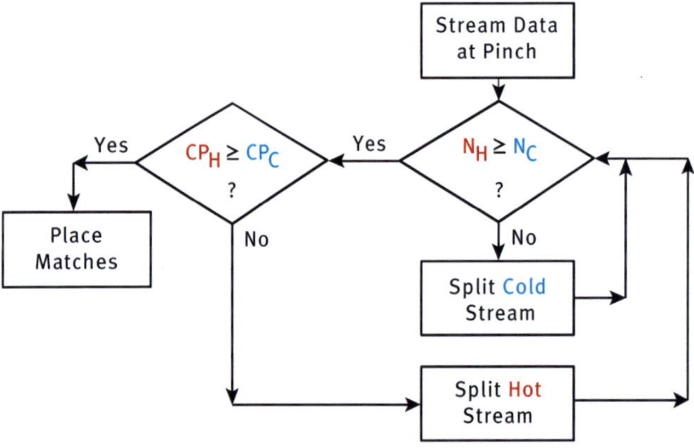

Fig. 3.19: Splitting procedure above the Pinch.

3.2.1.5 Network evolution

The HEN obtained using the design guidelines described previously is optimal with respect to its energy requirements, but it is usually away from the opimum total cost. Observing the Pinch division typically introduces loops into the final topology and leads to larger number of heat exchanger units. Thus, the final step in HEN design is evolution of the topology: identifying heat load loops and open heat load paths; then using them to optimise the network in terms of heat loads, heat transfer area, and topology. During this phase, formerly rigorous requirements – for example, that all temperature differences exceed ΔT_{min} and that cross-Pinch heat transfers be excluded – are usually relaxed.

The resulting optimisation formulations are typically nonlinear and involve structural decisions; consequently they are MINLP problems. Different approximations and simplifying assumptions can be introduced to obtain linear and/or continuous formulations. The design evolution step can even be performed manually by breaking the loops and reducing the number of heat exchangers.

Eliminating heat exchangers from the topology is done at the expense of shifting heat loads (from the eliminated Heat Recovery exchangers) to utility exchangers: heaters and coolers. Topology evolution terminates when the resulting energy cost increase exceeds the projected savings in capital costs, which corresponds to a total cost minimum.

Network evolution is performed by shifting loads within the network toward the end of eliminating excessive heat exchangers and/or reducing the effective heat transfer area. To shift loads, it is necessary to exploit the degrees of freedom provided by loops and utility paths. In this context, a loop is a circular closed path connecting two or more heat exchangers, and a utility path connects a hot with a cold utility or connects two utilities of the same type. Fig. 3.20 shows a HEN loop and a utility path. A network may contain many such loops and paths.

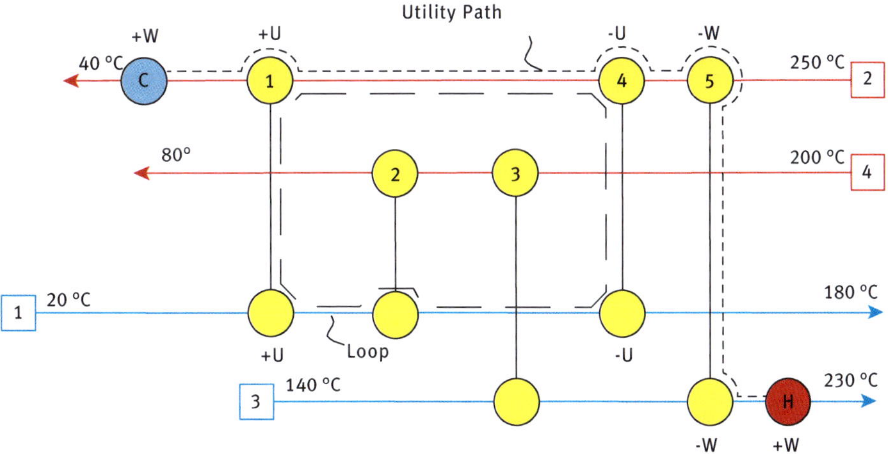

Fig. 3.20: Loop and path in a Heat Exchanger Network.

3.2.1.6 Working Session "HEN design for maximum Heat Recovery"
Assignment

The same flowsheet as in the working sessions in Chapter 2 "Setting targets and Heat Integration" is considered (Fig. 3.21, Table 3.2). The process stream data is the same (Table 3.2), so are the remaining specifications of the Heat Recovery problem (ΔT_{min} = 10 °C, Steam at 200 °C, CW at 25 °C). The assignment is to design a network of heat exchangers for Maximum Energy Recovery (MER).

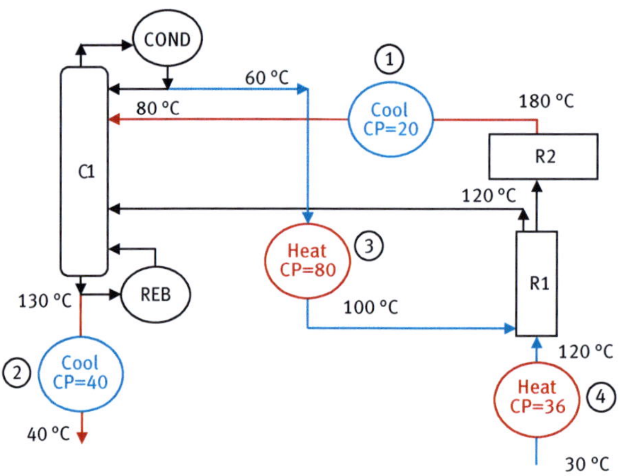

Fig. 3.21: Initial flowsheet for Working Session "Introduction to Heat Integration".

Tab. 3.2: Heat Recovery problem for Working Session "Introduction to Heat Integration".

Stream	T_S (°C)	T_T (°C)	CP (kW/°C)
1: Hot	180	80	20
2: Hot	130	40	40
3: Cold	60	100	80
4: Cold	30	120	36

Solution

The temperature boundaries are calculated first, to prepare the construction of the Problem Table. This step is illustrated in Fig. 3.22. Further, the Problem Table is constructed, filled in and calculated as shown in Fig. 3.23.

The heat cascade is obtained and made feasible as shown in Fig. 3.24. The cascade also provides all the needed energy targets.

Shifted Temperatures

Stream No and type	T_S [°C]	T_T [°C]	T^*_S [°C]	T^*_T [°C]
1 Hot	180	80	175	75
1 Hot	130	40	125	35
3 Cold	60	100	65	105
4 Cold	30	120	35	125

$\Delta T_{min} = 10\ °C$

Candidate Boundaries

175
125
125 ✕
105
75
65
35
35 ✕

Eliminate Duplicates

175
125
105
75
65
35

Fig. 3.22: Obtaining the temperature boundaries for Working Session "HEN design for maximum Heat Recovery".

Interval Temperature	Stream Population		$\Delta T_{interval}$	$\Sigma CP_H - \Sigma CP_C$	Heat Balance	Surplus / Deficit
175	1					
125	2	CP: 36	50	20	1,000	Surplus
105		CP: 80	20	24	480	Surplus
75			30	-56	-1,680	Deficit
65	CP: 20		10	-76	-760	Deficit
35		3	30	4	120	Surplus
	CP: 40	4				

Fig. 3.23: The Problem Table for Working Session "HEN design for maximum Heat Recovery".

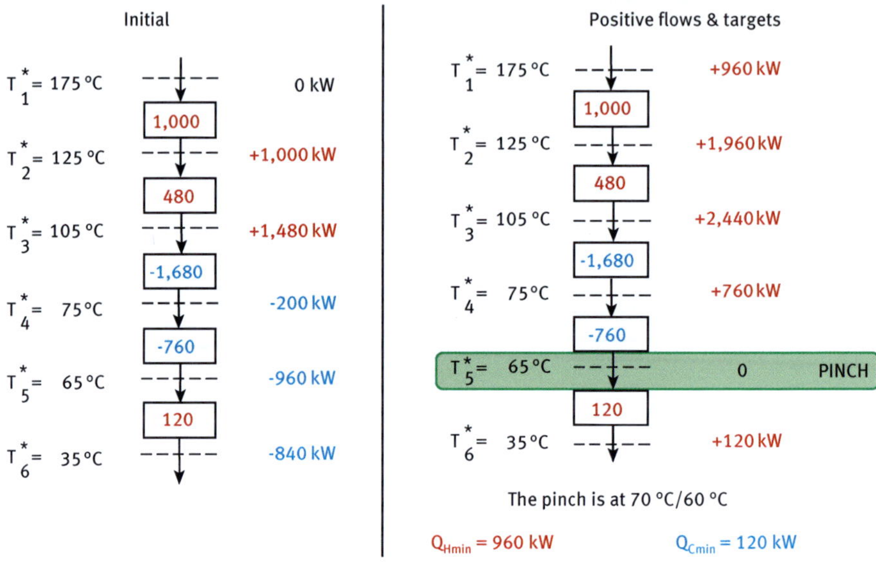

Fig. 3.24: The heat cascade for Working Session "HEN design for maximum Heat Recovery".

The targets are then used to sketch the empty design grid (Fig. 3.25).

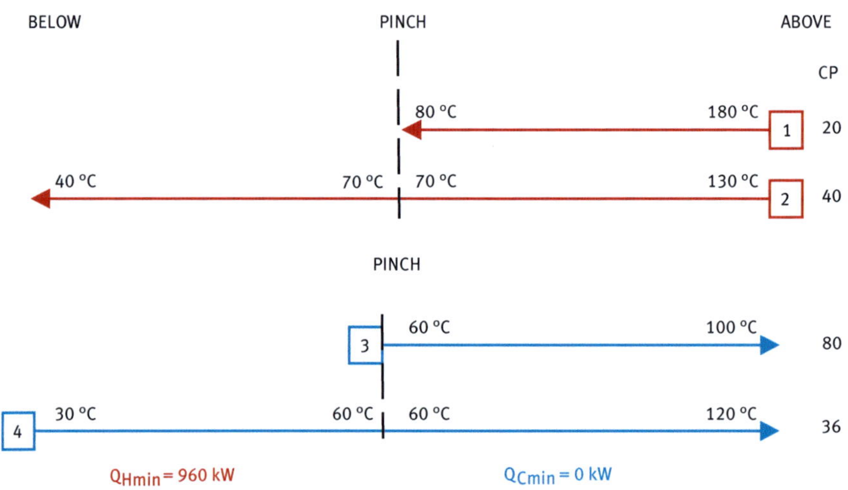

Fig. 3.25: The empty design grid for Working Session "HEN design for maximum Heat Recovery".

The design below the Pinch (Fig. 3.26) is simple, as there are only one hot and one cold stream. Since the cold stream (No. 4) has the smaller CP value, starting from the Pinch and moving to lower temperatures opens the T-H profiles, which makes the match feasible. Stream 4 also has the smaller duty and thus defines the load of the exchanger match. The residual cooling requirement on the hot stream (No. 2) is satisfied by a cooler with cooling water, with load of 120 kW.

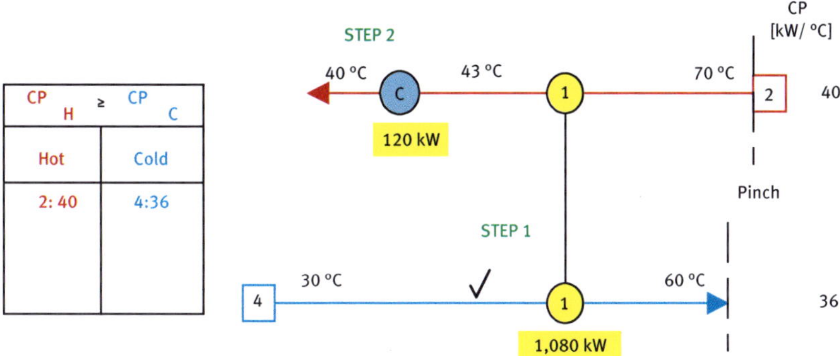

Fig. 3.26: Design below the Pinch for Working Session "HEN design for maximum Heat Recovery".

The design above the Pinch is more complicated (Fig. 3.27), as all four streams are present in that part. It takes several steps, as shown in that figure – first the recovery heat exchangers 2 and 3 are placed, after which follow the heaters H1 and H2. The sum of the heater duties is 960 kW, exactly equal to the minimum heating target.

The complete HEN design, merging the below and above-the-Pinch parts, is shown in Fig. 3.28. It is interesting to note that the design from Fig. 3.28 is not unique in terms of maximum Heat Recovery. It is also possible to obtain an alternative maximum Heat Recovery network, as shown in Fig. 3.29.

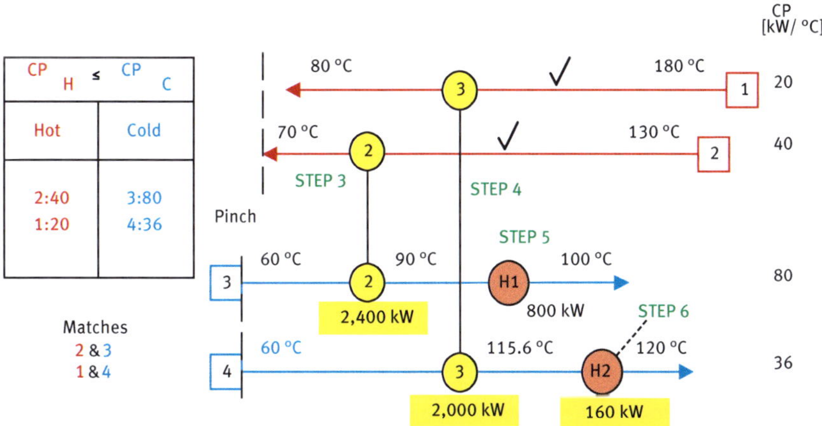

Fig. 3.27: Design above the Pinch for Working Session "HEN design for maximum Heat Recovery".

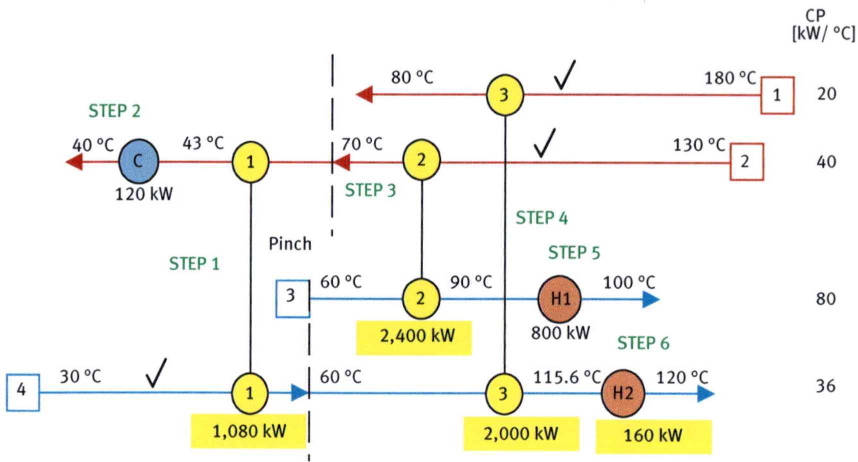

Fig. 3.28: The complete HEN design for Working Session "HEN design for maximum Heat Recovery".

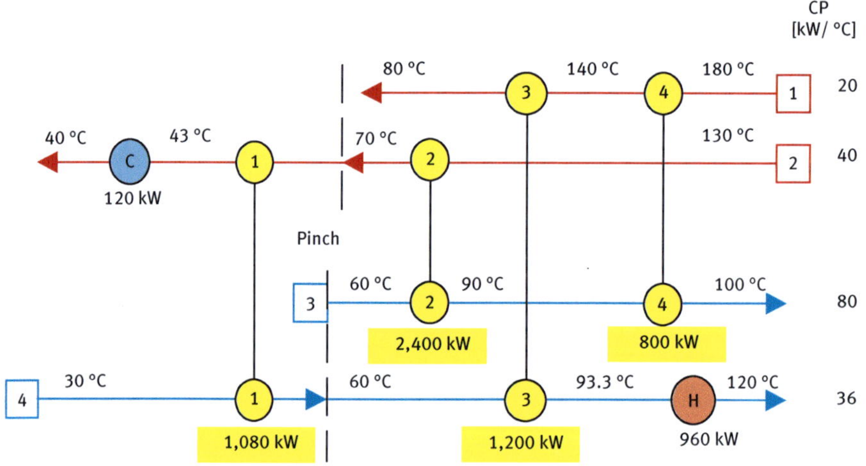

Fig. 3.29: An alternative HEN design for Working Session "HEN design for maximum Heat Recovery".

3.2.1.7 Working Session "An advanced HEN example"

Assignment

The stream data for a process are listed in Table 3.3. The hot utility is steam at 200 °C, the cold utility is water at 38 °C, and $\Delta T_{min} = 24\,°C$.

The following tasks are required to be performed:

(a) Plot the Composite Curves for this process.

(b) Determine the minimum heating requirement $Q_{H,min}$, the minimum cooling requirement $Q_{C,min}$, and the Pinch temperatures.

(c) Assuming that the cost of cooling water and steam is 18.13 and 37.78 $ kW^{-1} y^{-1}, plot the minimum annual cost for the utilities as a function of ΔT_{min} in the range 51–54 °C (in 1° increments).

(d) Design a network that features a minimum number of units and maximum energy recovery for $\Delta T_{min} = 24$ °C.

Tab. 3.3: Process stream data for Working Session "An advanced HEN example".

	Stream	T$_S$ (°C)	T$_T$ (°C)	CP (kW/ °C)
1	Cold	38	205	11
2	Cold	66	182	13
3	Cold	93	205	13
4	Hot	249	121	17
5	Hot	305	66	13

Solution

Answer to (a). The Composite Curves for the process stream data are shown in Fig. 3.30.

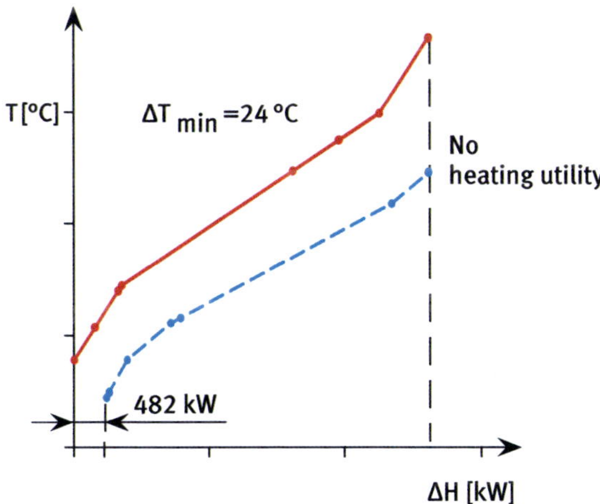

Fig. 3.30: Composite Curves for Working Session "An advanced HEN example".

Answer to (b). The position of the Composite Curves (CCs) in Fig. 3.30 indicates that, for $\Delta T_{min} = 24$ °C, this is a threshold problem at the cold end: $Q_{H,min} = 0$ kW. On the cold end there is an excess of hot streams, which means that some cold utility is required: $Q_{C,min} = 482$ kW.

Answer to (c). The utility cost plot, given in Fig. 3.31, is based on the data in Table 3.4.

Tab. 3.4: Utility requirements and costs for various ΔT_{min} values for Working Session "An advanced HEN example".

	ΔT_{min} (°C)			
	51	52	53	54
Q_H (kW)	0	0	27	57
Cost of heating ($/y)	0	0	1,020	2,153
Q_C (kW)	482	482	509	539
Cost of cooling ($/y)	8,739	8,739	9,228	9,772
Total utility cost ($/y)	8,739	8,739	10,248	11,926

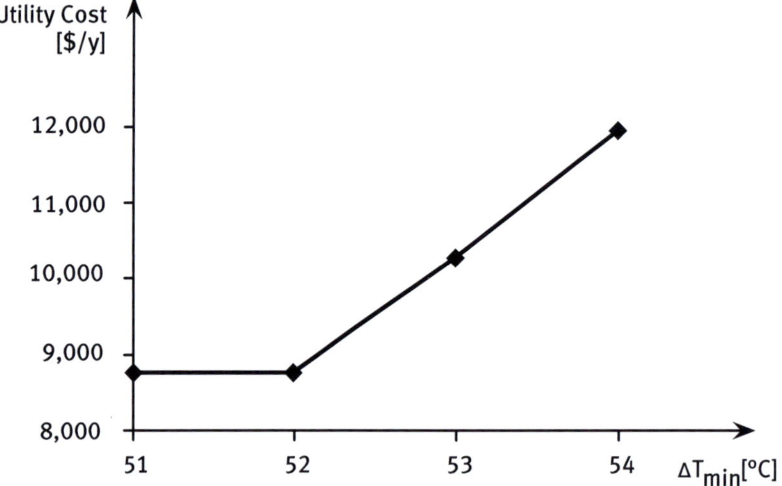

Fig. 3.31: Minimum allowed temperature difference ΔT_{min} versus annual utility cost for Working Session "An advanced HEN example".

Answer to (d). A pair of networks that feature the minimum number of units and maximum energy recovery is shown in Fig. 3.32. Please note that, besides being a threshold Problem, this is also one requiring stream splitting.

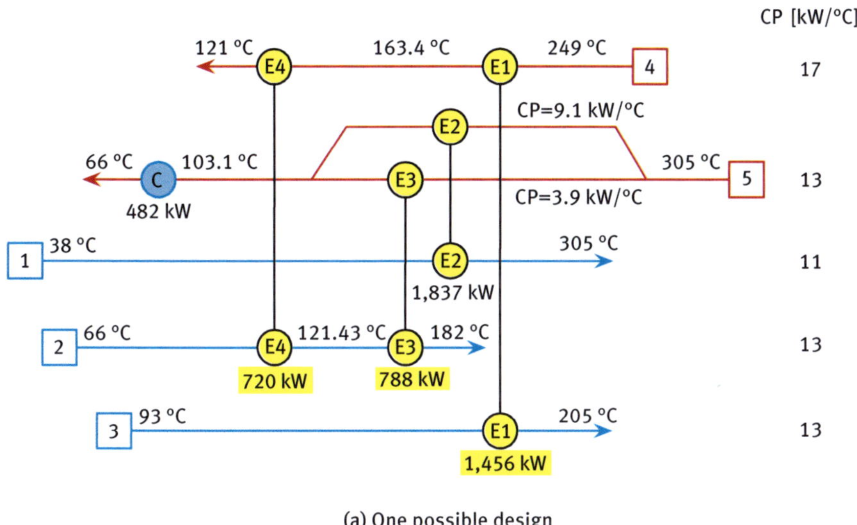

(a) One possible design

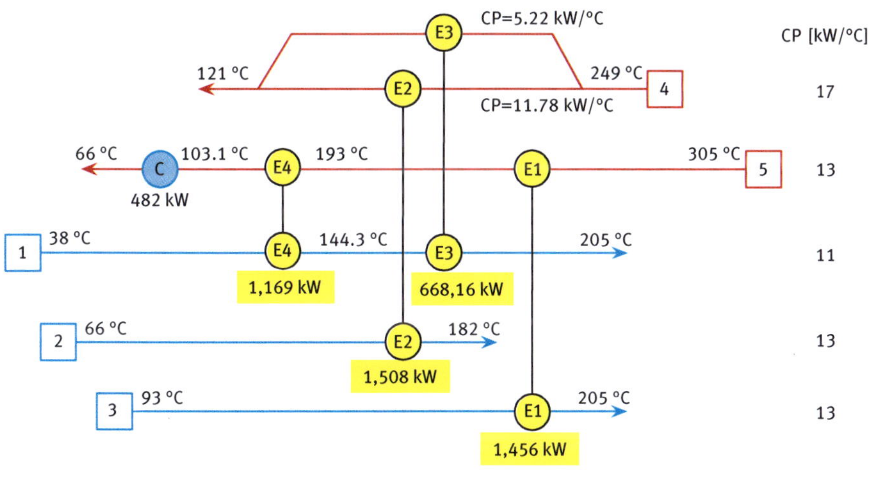

(a) A second possible solution

Fig. 3.32: Optimal Heat Exchanger Networks for Working Session "An advanced HEN example".

3.2.2 Methods using mathematical programming

3.2.2.1 MINLP approach

MINLP stands for "Mixed Integer Non-Linear Programming". This is one of the most used approaches for applying optimisation to chemical engineering and other industrial problems. As presented so far, the Pinch Design Method is based on a sequential strategy for the conceptual design of Heat Exchanger Networks. It first develops an understanding of the thermodynamic limitations imposed by the set of process streams, and then it exploits this knowledge to design a highly energy-efficient HEN. However, another approach has also been developed: the superstructure approach to HEN synthesis, which relies on developing a reducible structure (the superstructure) of the network under consideration and then reducing it using MINLP. An example of the spaghetti-type HEN superstructure fragments typically generated by such methods (Yee and Grossmann, 1990) is shown in Fig. 3.33.

This kind of superstructure is most often developed in stages – first formulated by Yee and Grossmann (1990) and more recently elaborated by Aaltola (2002) – or in blocks (Zhu et al., 1995; Zhu, 1997), each of which is a group of several consecutive enthalpy intervals. Zhu et al. (1995) investigated integrated HEN targeting and synthesis within the context of the so-called "block decomposition", which was further elaborated into an automated HEN synthesis method (Zhu, 1997).

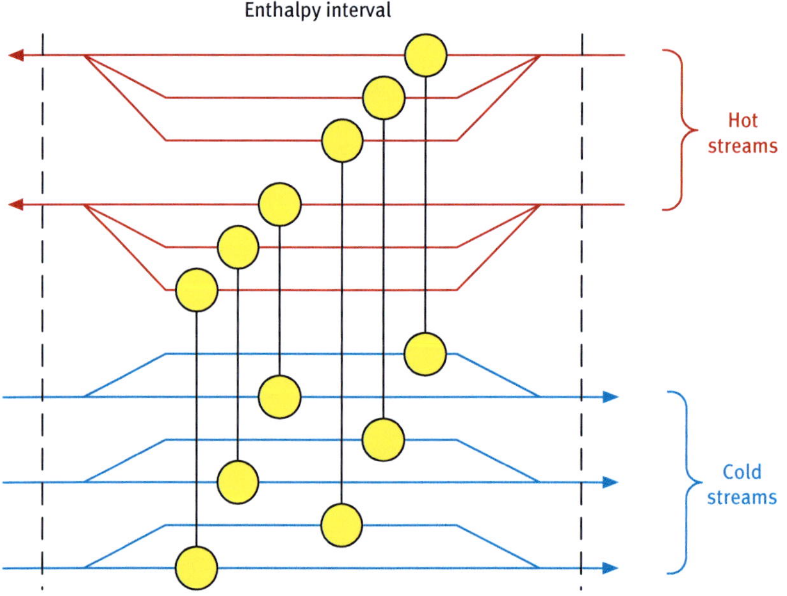

Fig. 3.33: Spaghetti superstructure fragment.

Within each block, each hot process stream is split into a number of branches corresponding to the number of the cold streams presented in the block and all cold streams are split similarly. This is followed by matching each hot branch with each cold branch.

Once developed, the superstructure is subjected to optimisation. The set of decision variables includes the existence of the different stream split branches and heat exchangers, the heat duties of the exchangers, and the split fractions or flow rates of the split streams. The objective function involves mainly the total annualised cost of the network. The objective function may be augmented by some penalty terms for dealing with difficult constraints. Because the optimisation procedure makes structural as well as operating decisions about the network being designed, it is called a structure-parameter optimisation.

"Structure-parameter optimisation" is a term widely used in Process Synthesis with a peculiar origin. It may be a little confusing to people familiar with mathematical optimisation only. In the context of optimisation, a parameter is an entity whose value is specified beforehand and is kept constant during the optimisation. However, in the early days of Process System Engineering, the term "parameter" was used more loosely referring to any properties of the modelled system – including variables, which could be represented by real numbers.

Depending on which assumptions are adopted, it is possible to obtain both MILP and MINLP formulations. Linear formulations are usually derived by assuming isothermal mixing of the split branches and then using piecewise linearisation of the heat exchanger capital cost functions.

Within the framework of the superstructure approach it is possible to include more than one type of heat exchangers – for example, direct heat transfer units (i.e., mixing) and further surface heat exchanger types (double-pipe, plate-fin, shell-and-tube). Soršak and Kravanja (2004) presented a method incorporating the different heat exchanger types into a superstructure block for each potential Heat Exchange match. Some other interesting works in this area are by Daichendt and Grossmann (1997) – using hierarchical decomposition, Zamora and Grossman (1998) – emphasising global MINLP optimisation, Björk and Westerlund (2002) exploring the non-isothermal mixing options, and Frausto-Hernández et al. (2003) – considering pressure drop effects.

3.2.2.2 A hybrid approach

It is clear that the superstructure methodology offers some advantages in the synthesis of process systems and, in particular, of Heat Exchanger Networks. Among these advantages are: (1) the capacity to evaluate a large number of structural and operating alternatives simultaneously; (2) the possibility of automating (to a high degree) the synthesis procedure; and (3) the ability to deal efficiently with many additional issues, such as different heat exchanger types and additional constraints (e.g., forbidden matches).

However, these advantages also give rise to certain weaknesses. First, the superstructure approaches, in general, cannot eliminate the inherent nonlinearity of the problem. Hence they resort to linearisation and simplifying assumptions, such as allowing only isothermal mixing of split streams. Second, the transparency and visualisation of the synthesis procedure are almost completely lost, excluding the engineer from the process. Third, the final network is merely given as an answer to the initial problem, and it is difficult to assess how good a solution it represents or whether a better solution is possible. Fourth, the difficulties of computation and interpreting the result grow dramatically with problem size. This is a consequence of the large number of discrete alternatives to be evaluated. Finally, the resulting networks often contain sub-networks that exhibit a spaghetti-type structure: a cluster of parallel branches on several hot and cold streams with multiple exchangers between them. Because of how superstructures are constructed, many such sub-networks usually cannot be eliminated by the solvers.

All these considerations highlight the fundamental trade-off between applying techniques that are based on thermodynamic insights, such as the Pinch Design Method, and relying on a superstructure approach. It would therefore be useful to follow a middle path taking the best from these two basic approaches.

One such middle path is the class of hybrid synthesis methods described next. A method of this class first applies Pinch Analysis to obtain a picture of the thermodynamic limitations of the problem; but then, instead of continuing on to direct synthesis, it builds a reduced superstructure. At this point the method follows the route of the classical superstructure approaches, including structure-parameter optimisation and topology simplifications. The cycle of optimisation and simplification is usually repeated several times before the final optimal network is obtained. All resulting networks feature a high degree of Heat Recovery, but not necessarily the maximum possible. A key component of this technique is avoiding the addition of unnecessary features to the superstructure, and this is an area where Pinch Analysis can prove helpful. A good example of a hybrid method for HEN synthesis is the block decomposition method (Zhu, 1997).

3.2.2.3 Comparison of HEN synthesis approaches

The networks obtained by the different synthesis methods have distinct features, which influence their total cost and their properties of operation and control. Because the Pinch Design Method incorporates the tick-off heuristic rule (Fig. 3.10), the networks synthesised by this method tend to have simple topologies with few stream splits. Both the tick-off rule and the Pinch principle dictate that utility exchangers be placed last, so they are usually located immediately before the target temperatures of the streams. However, the tick-off rule may also result in many process streams not having utility exchangers assigned to them, which reduces controllability. The Pinch Design Method may also reduce network flexibility because it relies on the Pinch decomposition of the problem (Fig. 3.6) and so, to a large degree, fixes the network behaviour.

Both the pure superstructure approach and the hybrid approach tend to produce more complex topologies. Their distinctive feature is the greater number of heat exchangers and stream splits, a result of how the initial superstructure is built. Spaghetti-type sub-networks also present a significant challenge to flexibility and control.

3.3 Grassroots and retrofits, impact of economic criteria

The HEN design method discussed in Section 3.2 refers to the design of new plants. It is useful for developing good understanding of the relevant heat and energy saving projects. However, most of the projects in industry try to make the most of existing facilities as opposed to completely replacing them with newly designed ones. Typically, they are geared towards improving the operation, removing bottlenecks, and improving efficiency with respect to energy and raw material utilisation.

The most commonly used term for such project types is *Retrofit*. Typical economic parameters or constraints are maximum allowed values for Payback Time and Investment Cost. The objective of a retrofit project is then to save as much energy as possible while satisfying these economic constraints. The strategy for retrofit problems needs to differ from that for new designs. Three different approaches seem to have been most popular historically (Kemp, 2007):

1. Synthesise a Maximum Energy Recovery (MER) – HEN for the existing plant, maximising the reuse of existing matches where possible. This implicitly puts the current HEN design as the procedure goal, therefore bound to produce poor results.
2. Start with the existing network and attempt to modify it step by step to conform closer to the MER targets. The procedure usually starts with taking the specifications of the current process streams, utilities and ΔT_{min}. Further, the MER targets are obtained by Pinch Analysis – minimum heating and cooling by various utilities, as well as Pinch locations. Further, the Grid Diagram is used to identify which heat transfer matches are the Pinch violators. Further, HEN topology modifications are proposed to correct the violations – addition of matches, shifting, removal.
3. Another option, reasonably attractive to industry, is to start from the existing HEN and identify the most critical changes to the network that would produce a substantial utility use reduction. This approach could be quite suitable if the current HEN is very far from the MER targets and the required topology modifications to achieve the targets would involve too high complexity and cost. Asante and Zhu (1997) showed that it is also highly effective in other situations. A key insight is the *Network Pinch*, which shows the current bottleneck to improving the energy recovery imposed by the current network structure.

This chapter provides next a short overview of the Network Pinch Method introduced by Asante and Zhu (1997).

3.3.1 Network optimisation

The *Network Pinch Method* is based on identifying the Heat Recovery bottleneck presented by the topology of the existing HEN. This is usually identified by performing optimisation of the HEN continuous properties – heat exchanger duties and stream temperatures, keeping the network topology fixed. In the course of this optimisation the ΔT_{min} constraint may eventually be relaxed even down to zero, leaving the total cost objective function to take care of identifying the best temperature differences for each heat exchanger.

3.3.2 The Network Pinch

Performing the mentioned optimisation on a HEN would eventually produce a combination of duties and temperatures, where at least one heat exchanger match will feature ΔT_{min} at some of its ends and the HEN utility demands are usually larger than the minima predicted by Pinch Analysis. This match is termed *the Pinching Match* and its end where the stream temperatures approach each other at ΔT_{min}, is termed the *Network Pinch*.

Consider the network in Fig. 3.34(a), which has $\Delta T_{min} = 10\,°C$. For the stream data, the targets are equal to 20 kW for hot utility and 60 kW for cold utility. There is a utility path from the heater **H**, through heat exchanger **E2**, to the cooler **C**. Exploiting the path, **E2** can be increased in size from 100 kW up to 112.5 kW, at which point the temperature difference at the cold end of **E2** becomes equal to the 10 °C constraint and the **E2** duty cannot be increased any further without violating the constraint (Fig. 3.34(b)). The load of the heater is 27.5 kW and that of the cooler is 67.5 kW, both 7.5 kW lower than the targets.

If ΔT_{min} is reduced to zero, a similar situation arises, where the targets are 0 kW for heating and 40 kW for cooling utility, presenting a threshold problem. For this case the network from Fig. 3.34 can be theoretically pushed to a little larger Heat Recovery with the heater load at 12.5 kW and cooler at 52.5 kW.

Thus the cold end of exchanger **E2** represents the Network Pinch. After identifying the Network Pinch, there are several types of actions that can be applied to improve the Heat Recovery. These are:
- **Resequence** matches. This means changing the order of matches, keeping the streams which they connect.
- **Repiping**. This action would change one of the streams in the match.
- **Addition of a new match**. This adds a new heat exchanger at a strategically chosen location in the existing network.
- **Stream splitting**. One of the streams in the considered match can be split, allowing a better adjustment of the CP of the stream branch remaining in the match.

For the considered case, it happens that adding a new match produces a network capable of achieving the Pinch targets (Fig. 3.35).

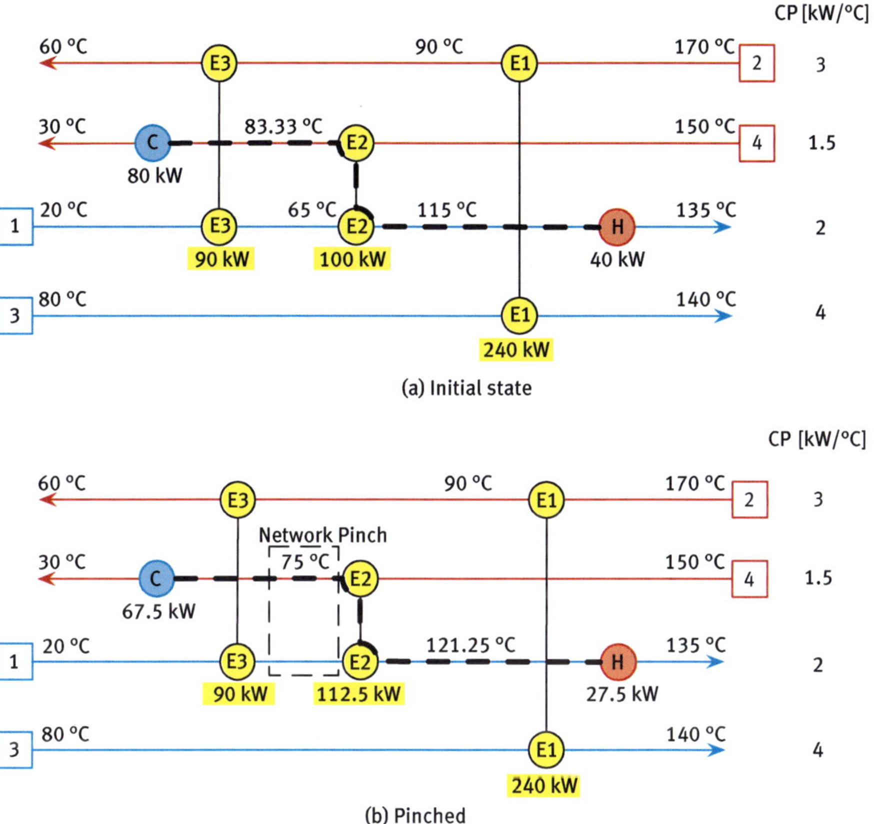

Fig. 3.34: An existing HEN.

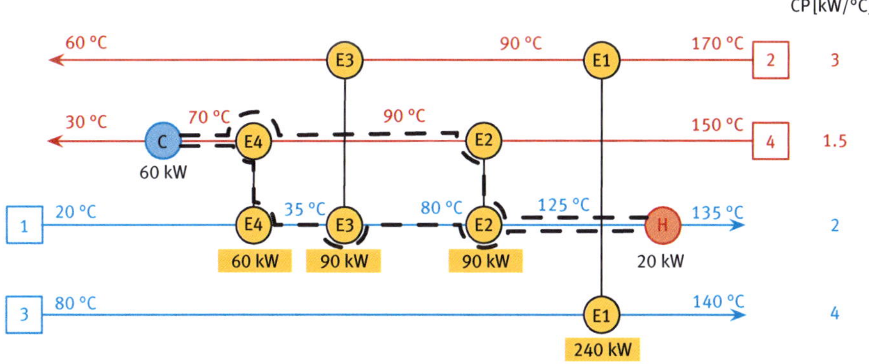

Fig. 3.35: A HEN with added heat exchanger (E4) achieving the targets.

Generally, after a retrofit step it may happen that Heat Recovery would improve, but more recovery would be still possible and even the Network Pinch location would most likely move to another match. In this case the procedure of identifying the Network Pinch followed by a new retrofit action can be applied. A more detailed instruction on the retrofit methodology of Network Pinch can be found in Asante and Zhu (1997).

3.4 Summary

In this chapter, the main concepts, representations, procedures and diagrams of HEN synthesis using the Pinch Design Method have been described and illustrated with working sessions. The most important topics covered in this chapter are:
- Grid Diagram for HEN synthesis and optimisation
- Pinch Decomposition
- Pinch Design Method with Matching Rules
- Design Evolution with Heat Load Loops and Paths
- Overview of other HEN synthesis methods and basics of retrofit.

References

Aaltola, J. (2002). Simultaneous synthesis of flexible heat exchanger network, *Applied Thermal Engineering*, 22, 907–918.

Anantharaman, R. (2011). *Energy efficiency in process plants with emphasis on heat exchanger networks – Optimization, thermodynamics and insight*, PhD Thesis, Norwegian University of Science and Technology, Trondheim, Norway.

Asante, N.D.K. and Zhu, X.X. (1997). An automated and interactive approach for heat exchanger network retrofit, *Chemical Engineering Research and Design*, 75(part A), 349–360.

Björk, K.M. and Westerlund, T. (2002). Global optimization of heat exchanger network synthesis problems with and without the isothermal mixing assumption, *Computers & Chemical Engineering*, 26(11), 1581–1593.

Bulatov, I. (2005). Retrofit optimization framework for compact heat exchangers, *Heat Transfer Engineerng*, 26, 4–14.

Daichendt, M.M. and Grossmann, I.E. (1997). Integration of hierarchical decomposition and mathematical programming for the synthesis of process flowsheets, *Computers & Chemical Engineering*, 22(1–2), 147–175.

Frausto-Hernández, S., Rico-Ramírez, V., Jiménez-Gutiérrez, A. and Hernández-Castro, S. (2003). MINLP synthesis of heat exchanger networks considering pressure drop effects, *Computers & Chemical Engineering*, 27(8–9), 1143–1152.

Furman, K.C. and Sahinidis, N.V. (2002). A critical review and annotated bibliography for heat exchanger network synthesis in the 20th century, *Ind. Eng. Chem. Res*, 41, 2335–2370.

Kemp, I.C. (2007). *Pinch Analysis and Process Integration. A User Guide on Process Integration for Efficient Use of Energy*, Amsterdam, The Netherlands: Elsevier (authors of the first edition: Linnhoff, B., Townsend, D.W., Boland, D., Hewitt, G.F., Thomas, B.E.A., Guy, A.R. and Marsland,

R.H. (1982 and 1994). A User Guide on Process Integration for the Efficient Use of Energy, Rugby, UK: IChemE).

Klemeš, J., Dhole, V. R., Raissi, K., Perry, S.J. and Puigjaner, L. (1997). Targeting and design methodology for reduction of fuel, power and CO_2 on Total Sites, *Applied Thermal Engineering*, 7, 993–1003.

Klemeš, J., Friedler, F., Bulatov, I. and Varbanov, P. (2010). *Sustainability in the Process Industry – Integration and Optimization*, New York: McGraw-Hill, 362 pp.

Klemeš, J. and Perry, S. (2007). Process optimisation to minimise energy use in food processing, in: Waldron K. (ed.), *Handbook of Waste Management and Co-product Recovery in Food Processing*, Vol. 1, pp. 59–89, Cambridge, UK: Woodhead.

Klemeš, J., Smith, R. and Kim, J.-K. (eds.) (2008). *Handbook of Water and Energy Management in Food Processing*, Cambridge, UK: Woodhead.

Linnhoff B. and Hindmarsh E. (1983). The Pinch design method for heat exchanger networks, *Chemical Engineering Science*, 38(5), 745–763.

Linnhoff, B., Townsend, D.W., Boland, D., Hewitt, G.F., Thomas, B.E.A., Guy, A.R. and Marsland, R.H. (1982). *A User Guide on Process Integration for the Efficient Use of Energy*, Rugby, UK: IChemE [revised edition published in 1994].

Linnhoff, B. and Vredeveld, D.R. (1984). Pinch technology has come of age, *Chemical Engineering Progress*, 80(7), 33–40.

Smith, R. (2005). *Chemical Process Design and Integration*, Chichester, UK: Wiley.

Smith, R., Klemeš, J., Tovazhnyansky, L.L., Kapustenko, P.A. and Uliev, L.M. (2000). *Foundations of heat processes integration*, Kharkiv, Ukraine: NTU KhPI (in Russian).

Soršak, A. and Kravanja, Z. (2004). MINLP retrofit of heat exchanger networks comprising different exchanger types, *Computers & Chemical Engineering*, 28(1–2), 235–251.

Smith, R., Jobson, M. and Chen., L. (2010). Recent developments in the retrofit of heat exchanger networks, *Applied Thermal Engng*, 30, 2281–2289.

Varbanov, P.S. and Klemeš, J. (2000). Rules for paths construction for HENs debottlenecking, *Applied Thermal Engineering*, 20, 1409–1420.

Wang, Y., Pan, M., Bulatov, I., Smith, R. and Kim, J.-K. (2012). Application of intensified heat transfer for the retrofit of heat exchanger networks, *Applied Energy*, 89, 45–59.

Yee, T.F. and Grossmann, I.E. (1990). Simultaneous optimization models for heat integration – II. Heat exchanger networks synthesis, *Comput. chem. Engng*, 14, 1165–1184.

Zamora, J.M. and Grossmann, I.E. (1998). A global MINLP optimization algorithm for the synthesis of heat exchanger networks with no stream splits, *Computers & Chemical Engineering*, 22(3), 367–384.

Zhu, X.X. (1997). Automated design method for heat exchanger networks using block decomposition and heuristic rules, *Computers & Chemical Engineering*, 21(10), 1095–1104.

Zhu, X.X., O'Neill, B.K., Roach, J.R. and Wood, R.M. (1995). Area-targeting methods for the direct synthesis of heat exchanger networks with unequal film coefficients, *Computers & Chemical Engineering*, 19(2), 223–239.

4 Total Site Integration

Maximising Heat Recovery at the process level is a good step toward better performance of industrial facilities. However, industrial processes rarely operate in isolation. They are usually organised in larger areas termed sites and are served by a common utility system. The processes interact with each other via the utility system. There are significant benefits to be gained from considering the complete sites as integrated energy systems, evaluating and optimising the energy generation, distribution, use and recovery. This chapter introduces a systematic framework and the associated algorithmic and visual tools for performing such an analysis.

This framework with its tools is widely known as Total Site Integration. Targeting for Total Sites is an extension of the established Pinch Technology Targeting methodologies and has been used extensively in industry. The Total Site targeting methodology includes Data Extraction methods, construction of Total Site Profiles, Total Site Composite Curves and the Site Utility Grand Composite Curve.

4.1 Introduction

Total Site Integration of industrial systems, based on the concept of the Site Heat Source and Heat Sink Profiles, was introduced by Dhole and Linnhoff (1993). Klemeš et al. (1997) made further advances in the field by adding targets for power cogeneration. The concept has been applied to industrial sites – in the beginning to refinery and petrochemical processes. These processes generally operate as parts of large sites or factories. The Heat Integration on such Total Sites is performed through a set of energy carriers – usually steam at several pressure levels, plus hot water and cooling water. The site processes are serviced by a central utility system providing the thermal energy carriers and power.

Total Site Integration has been applied to a number of chemical industrial sites (Matsuda et al., 2009) and even to a heterogeneous Total Site involving a brewery and several commercial energy users. It has also been used as the targeting stage for the synthesis of utility systems first in Shang and Kokossis (2004) and then elaborated for choosing steam header pressure levels by Varbanov et al. (2005).

The current chapter provides a guide through the basics of Total Site Integration. It starts with a definition, aided by description of the reasons for and the benefits from applying the concept. Further, the data extraction from process-level Heat Integration results is considered, which is one of the necessary steps for analysing new system designs. Next follow the steps of the basic Total Site Analysis – construction and use of the Total Site Profiles, the Total Site Composite Curves and the Site Utility Grand Composite Curve. The next section touches the subject of cogeneration targeting and its extended use for analysing the best power-to-heat ratio for a site. Advanced developments of the Total Site methodology are briefly reviewed after that – including more flexible specifications of the minimum allowed temperature differences, numerical tools for Total Site Heat Integration and Hybrid Power Systems. The chapter is concluded with a summary of the potential sources of further information.

The analysis of a Total Site almost invariably involves collection of information about the heating and cooling requirements on the site, thermodynamic analysis of these demands based on composite profiles and identification of the targets for minimum fuel use, power cogeneration and the amount of power import. The most frequently used procedure is shown in Fig. 4.1.

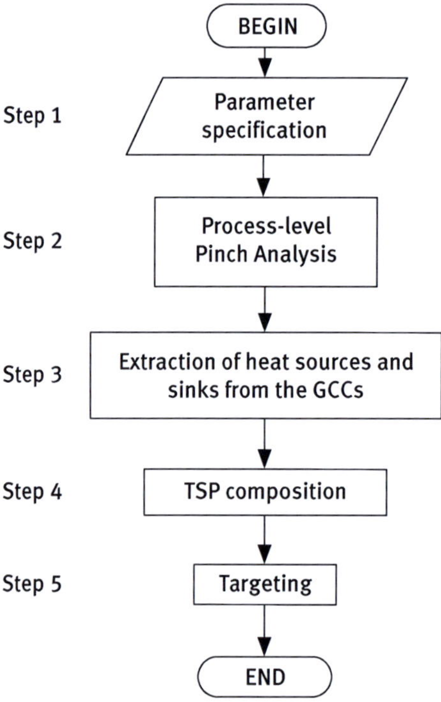

Fig. 4.1: Procedure for Total Site targeting.

4.2 What is a Total Site and what are the benefits?

It is rare to have single isolated industrial processes. The usual situation is to have clusters of processes, interlinked with one another by various lines for exchanging raw materials, side products, intermediates and, most importantly – energy carriers. The latter usually represent the most significant links.

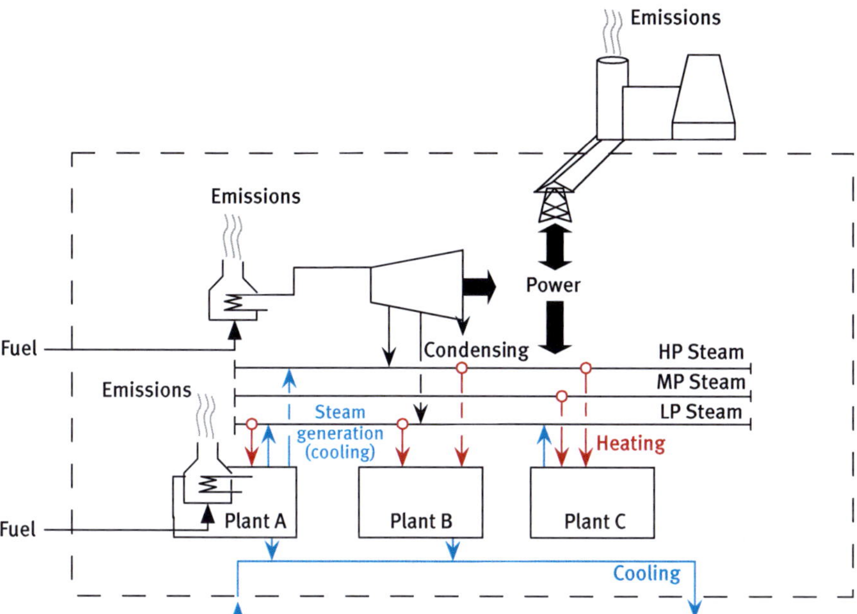

Fig. 4.2: Typical Total Site in industry.

4.2.1 Total Site definition

A typical chemical or other industrial site (Fig. 4.2) usually consists of a number of production and auxiliary processes. These processes require the supply of different utilities in order to carry out their functions. Such utilities are:

- **Process heating**. Steam is usually the preferred heating medium because of its high specific heat content in the form of latent heat and superheat. High temperature processes, however, may require heating with hot oil or directly with flue gas in furnaces.
- **Process cooling**. This is performed by using cooling water, ambient air, or refrigeration.
- **Power demands**. These arise from the need for driving process equipment – such as pumps, compressors, mills, etc. and also lighting and electric heating (where high precision and responsiveness is necessary).
- **Water supply and disposal**. This is also an important utility. It includes mainly the supply of freshwater, eventually treated to satisfy the water quality requirements, as well as the wastewater treatment, recycling and disposal.

The utility system is considered as supplying the site processes with heating and cooling utilities as well as satisfying their power demands. This follows from the two major components of Total Site Integration, closely related to each other: Heat Recovery and power cogeneration using the utility system.

In summary, an industrial site is a cluster of production processes, interlinked by a common utility system. There are many benefits of having the utilities provided by a central system. To a large extent a great benefit is the opportunity to share the large investment for the utility system between more processes, thus lowering the specific costs of utility supply.

Another advantage – a crucial one – is that a common utility system allows the processes to exchange utilities: the high-temperature cooling demand of a process may be used for utility (e.g. steam) generation instead of serving the demand with utility cooling. Lower temperature cooling demands in other processes can sometimes be served by the process-generated utility instead of applying boiler steam, utility hot oil or furnace flue gas. In this way, a central utility provides a kind of market place for exchanging utilities and enabling Heat Integration between separate processes, via intermediate energy carriers.

Having the common market place as an important degree of freedom in inter-process energy integration allows considering various options. When all processes on a site are considered together, the industrial site is termed a Total Site – implying that this is not just a collection of elements but an integrated system.

4.2.2 Total Site Analysis interfaces

To perform an analysis for inter-process energy integration, the Total Site methodology defines very clear interfaces. For analysing the Heat Exchange, the Heat Sinks and Heat Sources are used.

A Heat Sink is a representation of a heating demand. In this it is similar to a cold Process Stream. By the same analogy, it is defined by a starting temperature (T_S), a final/target temperature (T_T) and an enthalpy change (ΔH). Specifying these properties is significant to calculate all other information. In a similar way, a Heat Source is defined with a starting temperature (T_S), a final/target temperature (T_T) and an enthalpy change (ΔH), which is analogous to a hot Process Stream.

Heat Sources and Sinks can represent individual Process Streams – including residual heating or cooling demands – which process managers would like to expose to the exchange via the utility system. However, the reason for defining the separate concepts of Heat Sources and Sinks is to also provide the opportunity for extending process-level Heat Integration Analysis (i.e. Pinch Analysis) to the site level. In this way, Process Integration applies the known Information Technology principle of information encapsulation to provide more analytical flexibility to the site engineers.

4.3 HI extension for Total Sites: data extraction for Total Sites

When obtaining the Heat Sources and Sinks from process-level Pinch Analysis, the Grand Composite Curve (GCC) of each analysed process is used. GCC, as discussed above, is one of the representations of the process-utility interface for a single process. The GCC represents the remaining heating and cooling demands of the process after Heat Recovery has taken place within the process. The temperature scale of the GCC is in shifted temperature T*. The shifted temperatures are produced by shifting downwards (to lower values) the supply and target temperatures of hot streams by $0.5 \cdot \Delta T_{min}$, and shifting upwards (to higher values) the supply and target temperatures of the cold streams by $0.5 \cdot \Delta T_{min}$, building feasible Heat Exchange into the curve. The GCC shows the Pinch, the heating (Q_{Hmin}) and cooling (Q_{Cmin}) requirements to be supplied by external sources.

The individual GCCs can be used to identify the potential for inter-process Heat Recovery via steam mains. When a site houses several production processes, the GCC of each process may indicate steam levels suitable for the given process without guarantee that they will be suitable for the other processes. This suggests that trade-offs would be needed among the energy demands of the various processes on a site to minimise the overall utility demand.

Tab. 4.1: Stream specifications for Process A (Chemical Plant).

No.	Stream	Type	T_S (°C)	T_T (°C)	ΔH (MW)	CP (MW/°C)
1	A1	Hot	110	80	40.0	1.333
2	A2	Hot	150	149	180.0	180.000
3	A3	Cold	50	135	104.4	1.228
4	A4	Cold	85	100	82.3	5.487
5	A5	Cold	62	100	130.0	3.421
6	A6	Hot	92	55	130.0	7.647

4.3.1 The algorithm

The overall procedure for obtaining site-wide Heat Recovery targets is based on thermal profiles for the entire site – the Total Site Profiles (TSPs). Constructing the profiles requires data supplied in the form of Heat Sources and Heat Sinks (i.e. the input interfaces), plus specifications of the utilities – types and temperatures. Obtaining the data for the Heat Source and Sink specifications is referred to as "Total Site data extraction".

The data extraction procedure is shown in Fig. 4.3.

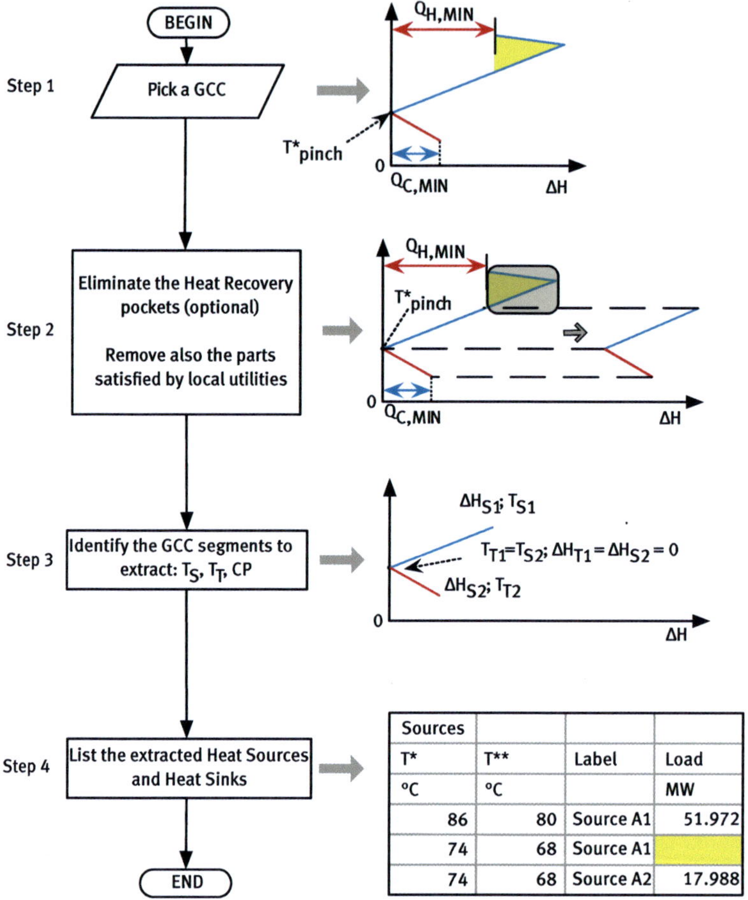

Fig. 4.3: Total Site data extraction procedure.

Tab. 4.2: Stream specifications for Process B (Food Processing Plant).

No.	Stream	Type	T_S (°C)	T_T (°C)	ΔH (MW)	CP (MW/°C)
1	B1	Hot	200	195	160.0	32.000
2	B2	Cold	20	54	10.0	0.294
3	B3	Cold	50	85	107.3	3.066
4	B4	Cold	100	120	130.0	6.500
5	B5	Hot	150	40	83.5	0.759
6	B6	Cold	80	95	48.0	3.200
7	B7	Hot	95	25	80.0	1.143

4.3.2 Step-by-step guidance

Consider a Total Site comprising four processes (Fig. 4.4, Tables 4.1–4.5). This will be labelled "Total Site Example 1". The minimum allowed temperature difference for all processes on the site is specified as $\Delta T_{min} = 12\,°C$.

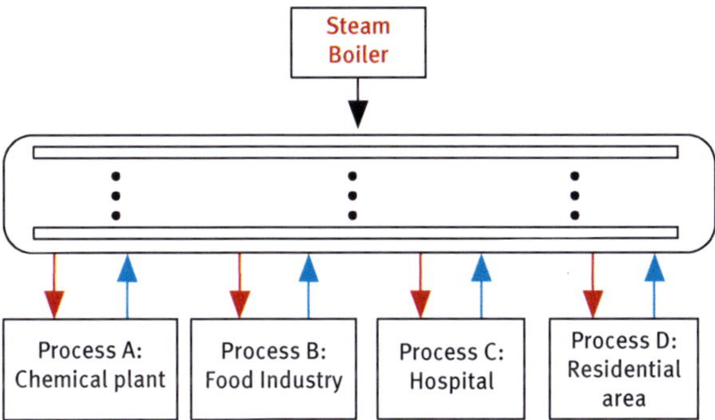

Fig. 4.4: A simple site.

Tab. 4.3: Stream specifications for Process C (Hospital).

No.	Stream	Type	T_S (°C)	T_T (°C)	ΔH (MW)	CP (MW/°C)
1	Soapy water	Hot	85	40	23.85	0.53
2	Condensate	Hot	80	40	96.4	2.41
3	Sanitary water	Cold	25	55	17.3	0.576
4	Laundry	Cold	55	85	18.0	0.600
5	BFW	Cold	33	60	12.0	0.446
6	Sanitary water	Cold	25	60	15.0	0.429
7	Sterilisation	Cold	82	121	34.1	0.874
8	Swimming pool water	Cold	25	28	23.1	7.700
9	Cooking	Cold	80	100	32.0	1.600
10	Heating	Cold	18	25	41.1	5.871
11	Bedpan washers	Cold	21	121	5.0	0.050

The extraction of Heat Sources and Heat Sinks from process GCCs are illustrated with the example of Process A from Fig. 4.4. Its Grand Composite Curve is shown in Fig. 4.5 with the heat cascade data as point coordinates. For information on how to perform Pinch Analysis and obtain the GCC, please refer to Chapter 2.

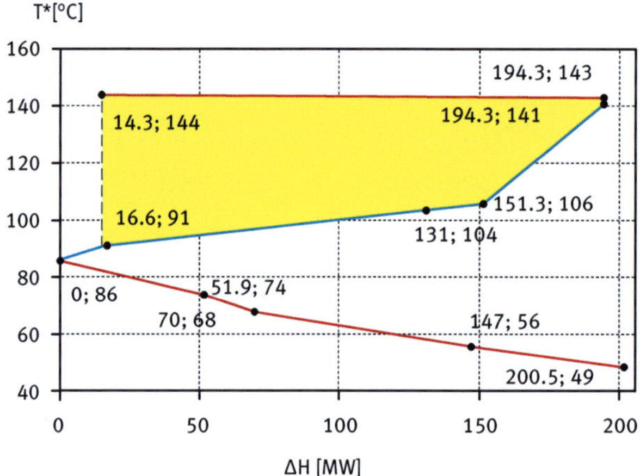

Fig. 4.5: GCC for Process A.

Tab. 4.4: Stream specifications for Process D (Residential Area).

No.	Stream	Type	T_S (°C)	T_T (°C)	ΔH (MW)	CP, (MW/°C)
1	Space heating	Cold	15	25	88.0	8.800
2	Hot water base	Cold	15	45	25.0	0.833
3	Hot water 2	Cold	15	45	65.0	2.167

Tab. 4.5: Available site utilities.

Type	Temperature range	Unit Cost ($/(kW · y))
Cooling Water	From 13 °C to 25 °C	150
LP Steam	150 °C	260
Hot Water	From 50 to 70 °C	–

The data in Fig. 4.5 are interpreted in the following way. The minimum heating demand of Process A (Q_{Hmin}) is 14.3 MW and the minimum cooling demand (Q_{Cmin}) is 200.5 MW. The Pinch is located at interval temperature T* = 86 °C, translating to 92 °C for the hot streams and 80 °C for the cold streams.

The next step (Step 2) of the extraction algorithm from Fig. 4.3 – the elimination of Heat Recovery pockets – is optional. Its idea is to eliminate the parts of GCC where additional intra-process Heat Recovery takes place and expose to inter-process exchange only the net heating or cooling demands. When Heat Recovery is the only goal, then pocket elimination results in minimising capital costs for the utility piping and utility heat exchangers and therefore is usual. The characteristic of "optional" for pocket elimination comes from the eventual additional goal of Total Site Integration – the cogeneration. If there exist GCC pockets with large temperature span and heat load (altogether – large area) and at sufficiently high temperatures, then it may be beneficial to export the GCC parts of the pockets. In such cases, steam at higher pressure may be generated and supplied by streams at the pocket top to the common utility system, in return the streams of the pocket bottom take lower pressure steam. In-between, the utility system can expand the higher pressure steam to the lower pressure, cogenerating power.

If extraction of the pocket segments is to be worthwhile, two conditions should be met:

- The pocket should span a considerable temperature difference to enable fitting a steam turbine in-between the upper and lower saturation temperatures defined by the pocket. In this way the pocket would be able to support generating a hot utility (steam) by the heat source in the pocket, and using a lower temperature hot utility (steam) to supply the cold sink.
- The amount of the heat source should be sufficiently large – larger than a few hundred kW, to justify the eventual additional investment cost of capital required to transform the heat source to a hot utility that can be used elsewhere on the Total Site and/or installing a steam turbine.

For the current discussion, it is assumed that the temperature difference for the pocket in Fig. 4.5 (between 143 °C and 106 °C) is insufficiently large to accommodate a steam turbine at a reasonable cost, and the pocket can be eliminated. The elimination is performed by first finding the vertical projection of the pocket top-left point of the GCC (in this section with coordinates $\Delta H = 14.271$ MW; T = 144 °C) onto the pocket bottom part of the curve. It is necessary to calculate the temperature at which the pocket bottom part of the GCC is at $\Delta H = 14.271$ MW. This operation can be performed by linear interpolation inside the corresponding GCC segment. It is illustrated in Fig. 4.6, where the unknown temperature is identified as $T^{*}_{\text{pocket bottom}} = 90.3$ °C. After this calculation, the GCC segments are removed from further consideration, which in Fig. 4.6 is denoted by greying that part of the GCC.

Next follows Step 3 of the extraction algorithm from Fig. 4.3 – identification of the GCC segments to extract. This operation involves reading the temperatures and ΔH values off the GCC or the related process heat cascade. This can be easily done by looking at Fig. 4.6.

Step 4 of the extraction algorithm is to list the extracted Heat Sources and Sinks. This also involves the additional calculation of the double-shifted temperature for each of the list entries (T^{**}). For a Heat Source, T^{**} is calculated by shifting its T^{*} values by $0.5 \cdot \Delta T_{\text{min}}$ downward (colder) and for a Heat Sink the shift is upward (hotter), which is completely analogous to the shifting for hot and cold Process Streams in process-level Pinch Analysis. After this calculation, the Heat Sinks are grouped into

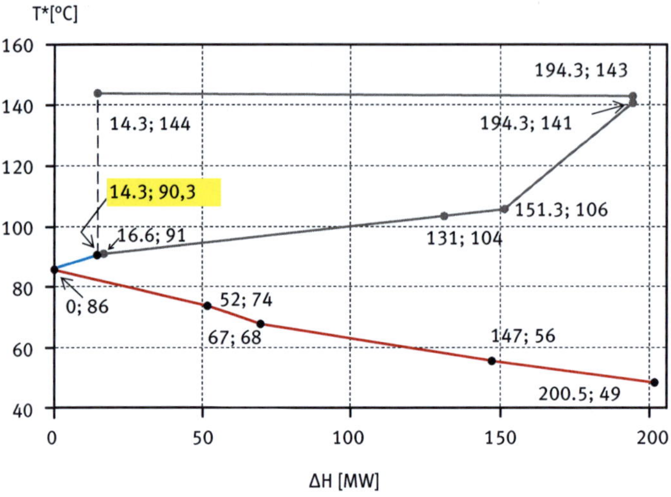

Fig. 4.6: Eliminating the pocket from the GCC of Process A.

one list and the Heat Sources into another. Both lists are sorted in descending order of the starting T** values.

Additional information can be indicated in these lists for better documentation and comprehension. For instance, since entries from many processes have to be listed in the same lists, the type and origin of the respective curve segment can be indicated. One useful option is to state that the curve segment is a Heat Source by adding the label "Source" and that it is the first Heat Source from Process A (A1), which yields a combined label "Source A1".

After applying the described actions to the GCC of Process A after pocket elimination, the resulting Heat Sources and Sinks are listed in Table 4.6 and Table 4.7.

Tab. 4.6: Heat Sources for Process A.

	T*	T**	Label	Load
	°C	°C		MW
Start	86	80	Source A1	51.972
End	74	68		
Start	74	68	Source A2	17.988
End	68	62		
Start	68	62	Source A3	77.028
End	56	50		
Start	56	50	Source A4	53.529
End	49	43		

Tab. 4.7: Heat Sinks for Process A.

	T*	T**	Label	Load
	°C	°C		MW
Start	86	92	Sink A1	14.271
End	90.3	96.3		

4.3.3 Working session

Assignment

The assignment for this exercise is for each of the Processes B, C and D to:

1. Perform process-level Pinch Analysis and draw the relevant Grand Composite Curves.
2. Identify site-level heat sources and heat sinks by extracting them from the process Grand Composite Curves. Note – remove the GCC pockets.
3. Calculate the double-shifted temperatures T^{**} of each Heat Source and Heat Sink and list the Heat Sources and Heat Sinks from Processes B, C and D, combined with those from Process A in two tables: one for Sources and another for Sinks.

Solution

The process-level Pinch Analysis has been performed for Processes B, C and D. The resulting GCC plots are shown in Figs. 4.7, 4.8, and 4.9.

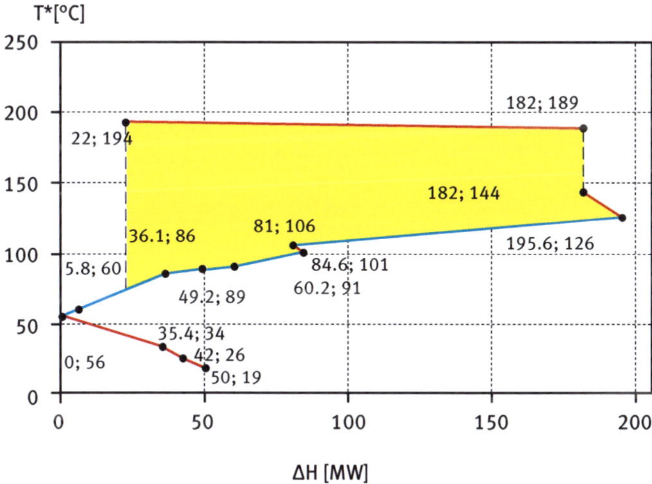

Fig. 4.7: GCC for Process B.

For Process B the minimum heating demand is Q_{Hmin} = 21.97 MW and the minimum cooling demand (Q_{Cmin}) is 50.17 MW. The Pinch is located at interval temperature T^* = 56 °C, translating to 62 °C for the hot streams and 50 °C for the cold streams. Processes C and D have no Pinch points and their targets can be read in the same way.

After removing the GCC pockets, the Heat Sources and Heat Sinks for Processes B, C and D have been identified and the T^{**} temperatures calculated. After combining these with the results from Process A, the full lists are given in Table 4.8 and Table 4.9.

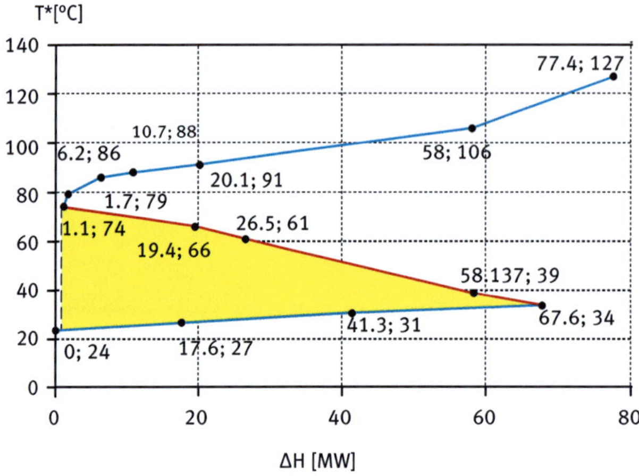

Fig. 4.8: GCC for Process C.

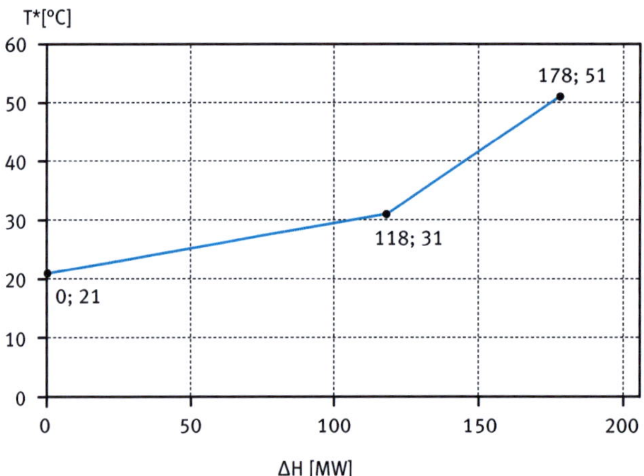

Fig. 4.9: GCC for Process D.

4.4 Total Site Profiles and Total Site Composite Curves

The next stage in the Total Site targeting procedure involves the combination of the Heat Sources from all site processes (A, B, C, D) into a composite Site Source Profile. The combination is performed among all Heat Sinks to produce the Site Sink Profile and among all Heat Sources to produce the Site Source Profile. The procedure is completely analogous to the one for constructing the Composite Curves in process-level Pinch Analysis (Chapter 2).

Tab. 4.8: Heat Sources for the example site.

	T*	T**	Label	Load
	°C	°C		MW
Start	86	80	Source A1	51.972
End	74	68		
Start	74	68	Source A2	17.988
End	68	62		
Start	68	62	Source A3	77.028
End	56	50		
Start	56	50	Source A4	53.529
End	49	43		
Start	56	50	Source B1	35.37
End	34	28		
Start	34	28	Source B2	6.8
End	26	20		
Start	26	20	Source B3	8
End	19	13		

Tab. 4.9: Heat Sinks for the example site.

	T*	T**	Label	Load
	°C	°C		MW
Start	86	92	Sink A1	14.271
End	90.3	96.3		
Start	56	62	Sink B1	5.83
End	60.0	66.0		
Start	60	66	Sink B2	16.14
End	73.9	79.9		
Start	24	30	Sink C1	1.084
End	24.2	30.2		
Start	74	80	Sink C2	0.6
End	79.0	85.0		
Start	79	85	Sink C3	4.55
End	86.0	92.0		
Start	86	92	Sink C4	4.5
End	88.0	94.0		
Start	88	94	Sink C5	9.372
End	91.0	97.0		
Start	91	97	Sink C6	37.86
End	106.0	112.0		
Start	106	112	Sink C7	19.404
End	127.0	133.0		
Start	21	27	Sink D1	118
End	31.0	37.0		
Start	31	37	Sink D2	60
End	51.0	57.0		

The procedure will be illustrated using two simple data sets termed "Total Site Example 2". Consider a simple site with two processes A and B. The specifications are:

- $\Delta T_{min} = 10\,°C$
- Boiler (High Pressure, HP) steam at saturation temperature $270\,°C$ (pressure 55.0 bar)
- Low Pressure (LP) steam at saturation temperature $140\,°C$ (pressure 3.6 bar)
- Cooling Water supplied at $15\,°C$ and returned to the utility system at $20\,°C$

The stream data for Process A are given in Table 4.10 and for Process B in Table 4.11.

Tab. 4.10: Total Site Example 2 – Process A streams.

Stream	Type	T_s (°C)	T_T (°C)	CP, (MW/°C)
A1	Cold	50	140	5
A2	Hot	100	30	6
A3	Cold	100	140	2

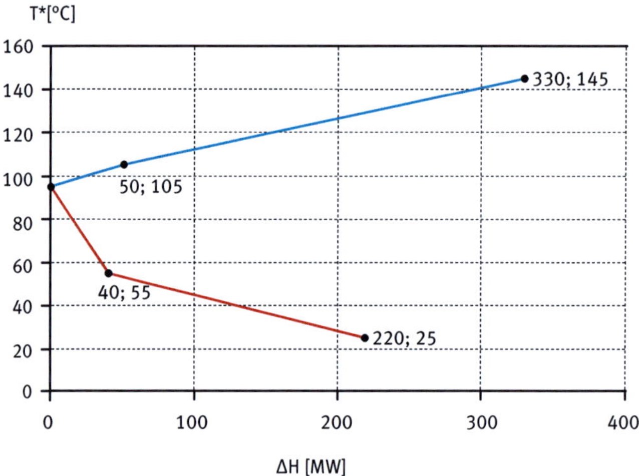

Fig. 4.10: Total Site Example 2 – GCC for Process A.

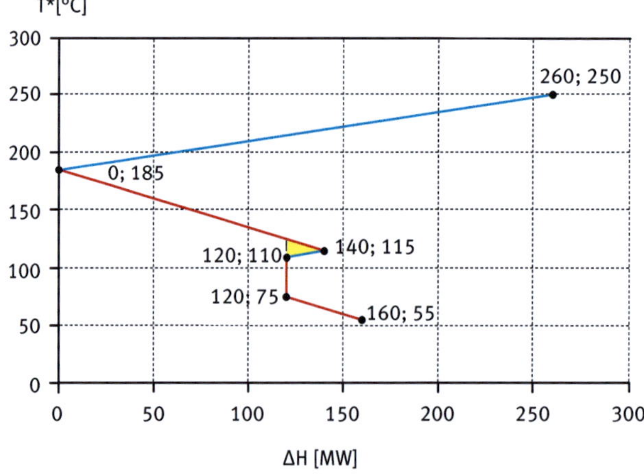

Fig. 4.11: Total Site Example 2 – GCC for Process B.

Tab. 4.11: Total Site Example 2 – Process B streams.

Stream	Type	T_S (°C)	T_T (°C)	CP, (MW/°C)
B1	Hot	190	120	6
B2	Cold	105	245	4
B3	Hot	80	60	2

The Heat Sources from both processes are listed in Table 4.12 and Table 4.13

Tab. 4.12: Total Site Example 2 – Heat Sources.

Segment	T^*_{start} (°C)	T^*_{end} (°C)	ΔH (MW)	T^{**}_{start} (°C)	T^{**}_{end} (°C)
Source A1	95	55	40	90	50
Source A2	55	25	180	50	20
Source B1	185	125	120	180	120
Source B2	75	55	40	70	50

Tab. 4.13: Total Site Example 2 – Heat Sinks.

Segment	T^*_{start} (°C)	T^*_{end} (°C)	ΔH (MW)	T^{**}_{start} (°C)	T^{**}_{end} (°C)
Sink A1	95	105	50.00	100.00	110.00
Sink A2	105	145	280.00	110.00	150.00
Sink B1	185	250	260.00	190.00	255.00

The composition of the Site Source Profile starts with identification of temperature intervals based on the T** scale and then mapping the Heat Sources from Table 4.12 to the intervals. Also, from the T** temperatures and the heat load of each segment, the corresponding CP values are calculated and added to the picture. The resulting diagram is shown in Fig. 4.12.

After that, for each interval in Fig. 4.12, the sum of the CP values of the Heat Sources present in it is taken and the interval enthalpy change calculated as the product of the latter sum and the temperature difference in the interval.

The data from Table 4.13 are also processed in a similar way and the diagram in Fig. 4.13 is obtained.

The next step is to build the Site Source Profile and the Site Sink Profile. For the Site Source Profile, the T** values temperature intervals in Fig. 4.12 are taken and listed in descending order forming the second column of Table 4.14 (T**, °C). In the first column of that table are calculated cascaded heat flows starting with zero and then subtracting the balance of each consecutive T** interval. This algorithm produces the data points for the Site Source Profile in such a way that it provides opportunity for steam generation at the highest possible pressure level (represented here by saturation temperature).

TB** °C	Source A1 CP MW / °C	Source A2 CP MW / °C	Source B1 CP MW / °C	Source B2 CP MW / °C	ΔT interval °C	ΣCP interval MW / °C	ΔH interval MW / °C
180							
			2		60	2	120
120							
					30	0	0
90							
	1				20	1	20
70							
	1			2	20	3	60
50							
		6			30	6	180
20							

Fig. 4.12: Combining the site Heat Sources (Total Site Example 2).

TB** °C	Sink A1 CP MW / °C	Sink A2 CP MW / °C	Sink B1 CP MW / °C	ΔT interval °C	ΣCP interval MW / °C	ΔH interval MW / °C
255						
			4.00	65	4	260
190						
				40	0	0
150						
		7.00		40	7	280
110						
	5.00			10	5	50
100						

Fig. 4.13: Combining the site Heat Sinks (Total Site Example 2).

Tab. 4.14: Data for the Site Source Profile. (Total Site Example 2)

ΔH, MW	T**, °C
0	180
−120	120
−120	90
−140	70
−200	50
−380	20

Tab. 4.15: Data for the Site Sink Profile. (Total Site Example 2)

ΔH, MW	T**, °C
0	100
50	110
330	150
330	190
590	255

In a similar, but mirror way, the T** values temperature intervals in Fig. 4.13 are listed in ascending order forming the second column of Table 4.15 (T**, °C) and the ΔH column is filled with cascaded heat flow values from top to bottom (from lower to higher temperature).

The next step is to plot the data points from Table 4.14 and Table 4.15, which yields the pair of Total Site Profiles: Site Source Profile and Site Sink Profile (Fig. 4.14).

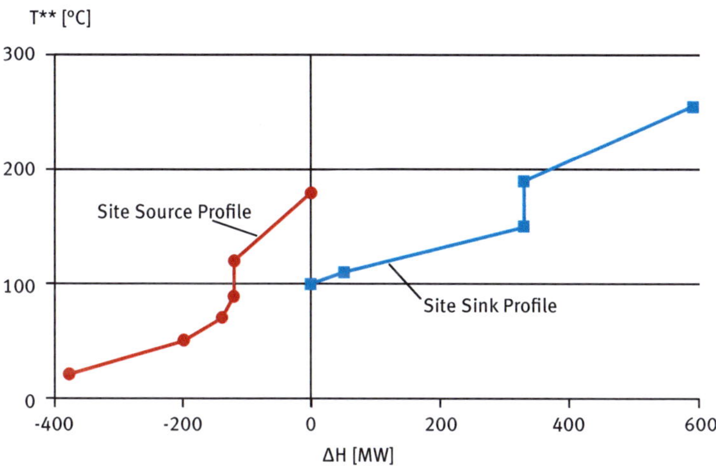

Fig. 4.14: Total Site Profiles (Total Site Example 2).

The evaluation so far has revealed that the total heating requirement of the Site Heat Sinks is 590 MW and the total cooling requirement of the Site Heat Sources is 380 MW. Without any inter-process Heat Recovery these would be the final amounts requested from the utility boiler and from the cooling water. Inter-process Heat Recovery can be performed using a utility as a carrier. It is important to underline that such an evaluation requires that the temperature levels of the utilities are specified before executing the procedure.

From all site utilities, the ones intersecting both Total Site Profiles can be used for such Heat Recovery. For the Total Site Example 2, it can be seen that only LP steam intersects both the Heat Sink and Source Profiles, as illustrated in Fig. 4.15. So, for the example under consideration, the only vehicle for inter-process Heat Recovery is the LP steam.

The maximum possible Heat Recovery through a utility system can be targeted by using the Site Sink Profile and the Site Source Profile in combination with the steam header saturation temperatures. Site-level Composite Curves for utility generation and use are constructed, accounting for feasible heat transfer.

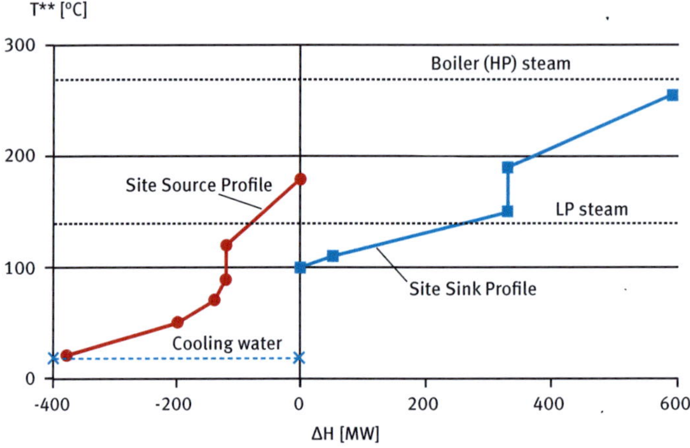

Fig. 4.15: Total Site Profiles with utilities (Total Site Example 2).

To understand the set of site-level curves, it is necessary to describe how the process-utility interface functions. Process heat sources can be used to generate utilities – in most cases steam at various levels. This is a heat transfer stage. The generated utility is represented by the Hot Utility Generation Composite Curve (HUGCC), also known as Site Source Composite Curve from previous sources – e.g. the pivotal Total Site work by Klemeš et al. (1997) also summarised in a recent book (Klemeš et al., 2010). At the next stage, the generated utility is supplied to a system of vessels usually termed a "main" or a "header". The header system then routes the utility flows to the potential users. There the utility can be directly consumed – e.g. live steam used in reactors and distillation columns. More often, however, the utility is used for process heating via another heat transfer stage. The supply of heating utility from headers is represented by the Hot Utility Use Composite Curve (HUUCC) also known as Site Sink Composite Curve (Klemeš et al., 1997, 2010).

The two heat transfer stages described above imply the need to ensure minimum temperature differences between the Site Source Profile and the HUGCC as well as between the HUUCC and the Site Sink Profile, via the double shifting of the process Heat Source and Sink temperatures to the T** scale, as discussed previously. The two site CCs are analogous to the individual process CCs. The Source CC is built starting from the highest possible steam level. The steam generation at each level is maximised before the next lower temperature levels are considered. This ensures maximum utilisation of the temperature potential of the Heat Sources. The remainder of the Site Heat Source Profile, that does not overlap the Source CC, is served by cooling water. Building the Sink CC follows a symmetrical procedure, starting from the lowest possible steam level. The utility use of this level is maximised before moving up to the next higher temperature level and so on until steam with the highest possible level is used – including boiler-generated steam.

The construction of the Site CC for Example 2 is shown in Fig. 4.16. Let us start with the Utility Generation CC (Fig. 4.16a). At Step 1 it can be seen that no HP steam can be generated by the Site Heat Sources and therefore the curve stays at ΔH of zero MW. Moving toward the LP steam level (Step 2), it is obvious that some LP steam can be generated. The exact amount is identified by intersecting the LP steam temperature level with the Site Heat Source Profile (Step 3) and this is equal to 80 MW. The remaining Heat Sources have too low a temperature at the current utility specifications and therefore are served by 300 MW cooling water. The construction of the Utility Use Composite Curve is made in a similar way, identifying that the LP steam demand is 260 MW and the genuine HP steam demand is 330 MW.

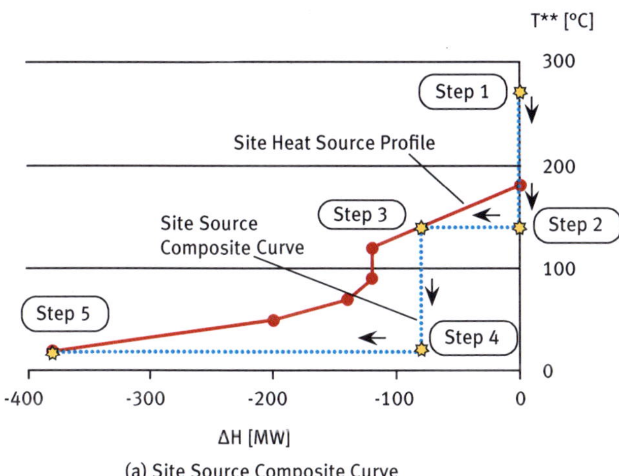

(a) Site Source Composite Curve

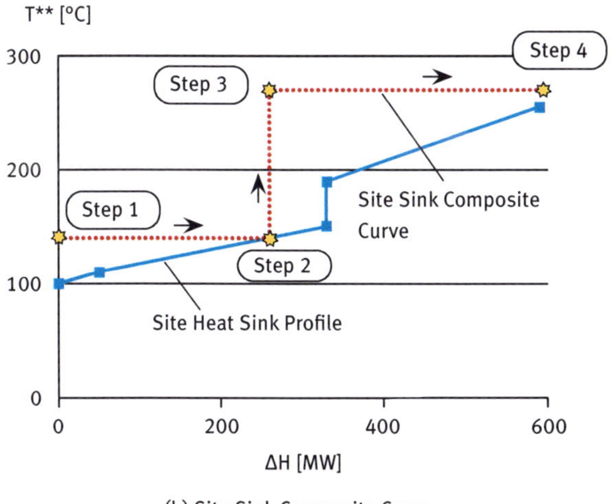

(b) Site Sink Composite Curve

Fig. 4.16: Construction of Site Source CC and Site Sink CC (Total Site Example 2).

After the construction of the Site CC, to target the site Heat Recovery via the intermediate utilities, the Source CC (Utility Generation CC) is shifted towards the Sink CC (Utility Use CC) to overlap, much in the same way as for the Composite Curves in process-level Pinch Analysis, thereby illustrating the Total Site Heat Recovery possible through the steam system. The amount of Heat Recovery for the Total Site is indicated by the amount of overlap between the CCs. Hence, Heat Recovery is maximised when the two curves touch and cannot be shifted further. The area where the curves touch, which is usually confined between two steam levels, is the Total Site Pinch. It is characterised by opposite utility load balances above and below. Above the Total Site Pinch there is a net deficit of heat, and below it is a net excess of heat. For Example 2, the shift of the Total Site Composite Curves is illustrated in Fig. 4.17. There at the LP steam level there is 80 MW LP steam generation resulting from the Site Source CC and 260 MW LP steam use resulting from the Site Sink CC, making a net deficit of 180 MW of LP steam to be cascaded from the HP steam (boiler).

At the lower utility level – that of the cooling water, there is a net excess of heat coming entirely from the Heat Source Profile totalling 300 MW. As a result, the Total Site Pinch for Example 2 is located at the utility levels LP steam – cooling water.

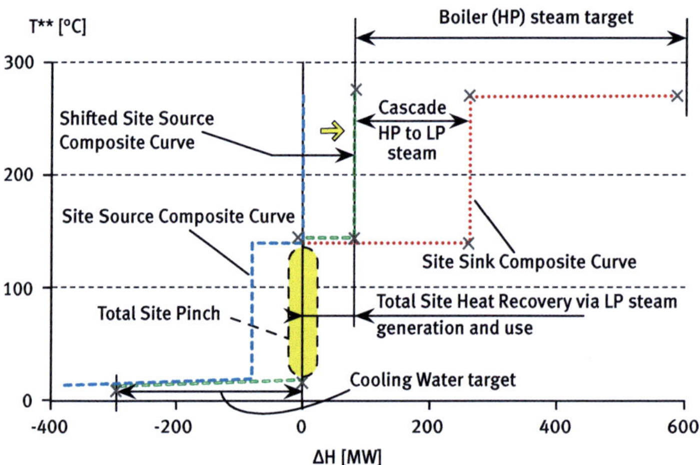

Fig. 4.17: Shift of the HUGCC toward the HUUCC and identification of the Total Site Pinch (Total Site Example 2).

4.5 Site Utility Grand Composite Curve (SUGCC)

Another targeting tool available for Total Sites is the Site Utility Grand Composite Curve (SUGCC), which evaluates the potential cogeneration that is available in the Site Utility system due to the distribution of steam to the site process Heat Sinks and the generation of steam from site boilers and process heat sources. The SUGCC can be derived from the targeting information already obtained in the form of the Total Site Profiles and Total Site Composite Curves.

Visually, the SUGCC can be obtained geometrically by subtracting the segments of the Site Sink CC from those of the Site Source CC. The algorithm of constructing the SUGCC includes:

1. Plot the highest temperature utility from zero ΔH to the right (data source: Site Source CC).
2. At the temperature reached in Step 1, plot to the left the utility use at the same utility level (data source: Site Sink CC). If the current utility is steam, its generation and use will have the same temperatures.
3. If there is a lower temperature utility, continue the curve vertically down to the next lower temperature utility level and select it as the current one. The SUGCC ends when there are no other lower temperature utilities. The vertical line is drawn for continuity of the curve and does not represent any heat transfer directly.
4. Repeat the above steps for the new "current" utility level.

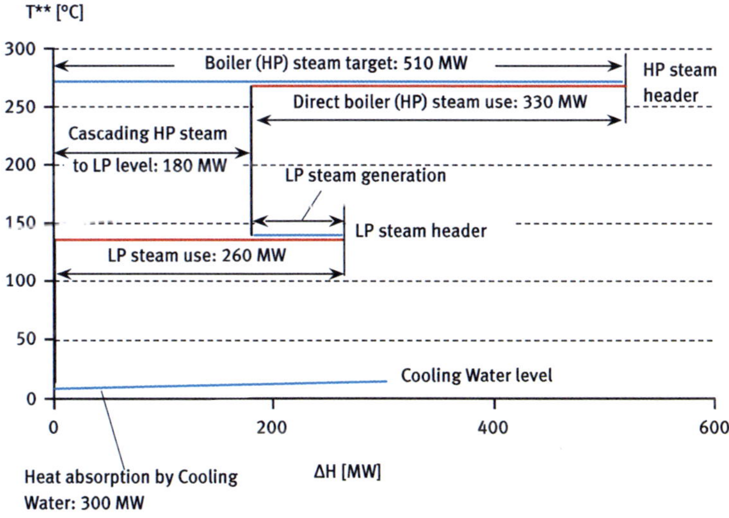

Fig. 4.18: The Site Utility Grand Composite Curve (SUGCC) – Total Site Example 2.

Fig. 4.18 illustrates the described procedure and its result when applied to the case of Total Site Example 2.

4.6 Combined Heat and Power Generation (CHP, Cogeneration) targeting

One can see from the SUGCC in Fig. 4.18 that between the HP steam level and the LP steam level there is an enclosed area. This represents the cogeneration potential if a higher pressure steam is expanded in a turbine to a lower pressure level. This insight has been documented FIRST by Raissi (1994) and further developed into a complete methodology in Klemeš et al. (1997). Klemeš et al. (1997) defined a simple proportionality coefficient, whose value is usually evaluated for each industrial site separately. This cogeneration targeting model is referred to as "the T-H model" because it is based on heat flows through the steam system.

Using SUGCCs allowed Klemeš et al. (1997) to set thermodynamic targets for cogeneration along with targets for site-scope Heat Recovery minimising the cost of utilities. Satisfying the goal of maximum Heat Recovery leads to a minimum boiler Very High Pressure (VHP) steam requirement, which in turn can be achieved by maximising steam recovery. At such condition the power generation by steam turbines is also minimal, which has the effect of maximising imported power. This scenario can be represented by the site CCs that are shifted to a position of maximum overlap (i.e., pinched). This target represents the thermodynamic limitation on system efficiency, but this is not a specification that must be achieved. The case of minimising the cost of utilities is handled by exploring the trade-off between steam recovery and cogeneration potential by steam turbines. If design guidelines are thus based on minimising cost, then the resulting utility network design is usually different from that produced when aiming for minimum fuel consumption.

Mavromatis and Kokossis (1998) proposed a simple model of back-pressure steam turbine performance. In this model, the performance of a steam turbine is related to its size (in terms of maximum shaft power) and to part-load performance. The shaft power is modelled as a function of the steam mass flow, a function that is known as the Willan's line. This model was later extended to condensing steam turbines by Shang (2000). All these works follow the same model structure and employ the same equations; however, they use different values for the turbine regression coefficients.

The intercept of the Willan's line was mapped by Mavromatis and Kokossis (1998), as well as by Shang (2000) as identical to the turbine energy losses, also assuming a fixed loss rate. Varbanov et al. (2004a) introduced improvements to those models by: (1) recognising that the Willan's line intercept has no direct physical meaning and is simply the intercept of a linearisation; and (2) accounting for both inlet and outlet pressures of the steam turbines. The improved steam turbine models have been incorporated into methodologies for simulating and optimising steam networks and also have been used to target the cogeneration potential by assuming a single large steam turbine for each expansion zone between two consecutive steam headers (Varbanov et al., 2004b).

In the example in Fig. 4.18 from the total 510 MW HP steam generated by the boilers, 330 MW is sent for process use and 180 MW is expanded to the LP steam level. If this amount of steam is expanded in a steam turbine, then this will produce power that can be used on the site.

4.6.1 A simple cogeneration model

One adequate cogeneration model suitable for targeting based on the SUGCC information has been presented by Varbanov et al. (2004a). This section provides a summary of the model.

The steam turbine performance at any load is estimated using the popular Willan's line:

$$W = n_{ST} \cdot m_{STEAM} - W_{INT}. \tag{4.1}$$

In the case of Total Site targeting, the current steam turbine load is equal to the maximum turbine load and Equation (4.1) takes the following form:

$$W_{MAX} = n_{ST} \cdot m_{max} - W_{INT}. \tag{4.2}$$

The Willan's line slope n_{ST} is estimated based on the isentropic enthalpy drop across the turbine and the maximum steam flowrate:

$$n_{ST} = \frac{L+1}{B_{ST}} \cdot \left(\Delta h_{is} - \frac{A_{ST}}{m_{max}} \right). \tag{4.3}$$

In Equation (4.3), L is referred to as the intercept ratio, A_{ST} and B_{ST} are regression parameters. The Willan's line intercept is:

$$W_{int} = \frac{L}{B_{ST}} \cdot \left(\Delta h_{is} \cdot m_{max} - A_{ST} \right). \tag{4.4}$$

The steam turbine parameters to A_{ST} and B_{ST} used in Equations (4.3) and (4.4) are calculated from the following regression relationships:

$$A_{ST} = b_0 + b_1 \cdot \Delta T_{sat} \tag{4.5}$$

$$B_{ST} = b_2 + b_3 \cdot \Delta T_{sat}. \tag{4.6}$$

The intercept ratio L from Equations (4.3) and (4.4) has also been found to depend on the saturation temperature drop across steam turbines:

$$L = a_L + b_L \cdot \Delta T_{sat}. \tag{4.7}$$

The total energy extracted from the expanding steam can be estimated using a fixed specification for the mechanical efficiency of the steam turbine and generator:

$$W_{total} = \frac{W}{\eta_{mech}}. \tag{4.8}$$

Alternatively, the evaluation can be done from a linear regression model if regression data are available:

$$W_{total} = a_{tot} + b_{tot} \cdot W. \tag{4.9}$$

The latter allows estimation of the enthalpy of the turbine exhaust as follows:

$$h_{STEX} = h_{ST,In} - \frac{W_{total}}{m_{STEAM}}. \tag{4.10}$$

The values of the performance regression parameters derived in (Varbanov et al., 2004a), for a given set of steam turbines, are listed in Table 4.16. Additionally, further simplifications of this model may be used, where some of the calculated coefficients are fixed based on additional sensitivity analysis.

Tab. 4.16: Regression coefficients for turbine performance estimation.

Coefficient	Unit	Value
b_0	MW	$-2.080 \cdot 10^{-8}$
b_1	MW/°C	0.000297
b_2	–	1.60200
b_3	1/°C	−0.001600
a_L	–	−0.01
b_L	1/°C	0.000326
a_{tot}	MW	0.1422
b_{tot}	MW/°C	1.0174

4.6.2 Targeting CHP using the SUGCC

The simple model from the previous section can be used for targeting the cogeneration potential of Total Site Utility Systems. This is performed by representing the Total Site by the corresponding SUGCC. In each expansion zone, where some non-zero area is enclosed by the SUGCC, a single simple steam turbine is assumed; the heat flow at the entrance of the expansion zone is estimated from the net cascaded heat flow and the steam level parameters. The model is further applied and the power cogeneration target estimated.

For Total Site Example 2, the steam turbine parameters to use are listed in Table 4.17.

Tab. 4.17: Steam turbine regression coefficients for Total Site Example 2.

Parameter	Unit	Value
b_0	MW	0.5
b_1	MW/°C	0.008
b_2	–	1.18
b_3	1/°C	0.03
L	–	0.15
η_{mech}	–	0.95

The cogeneration estimation based on these data is shown in Fig. 4.19.

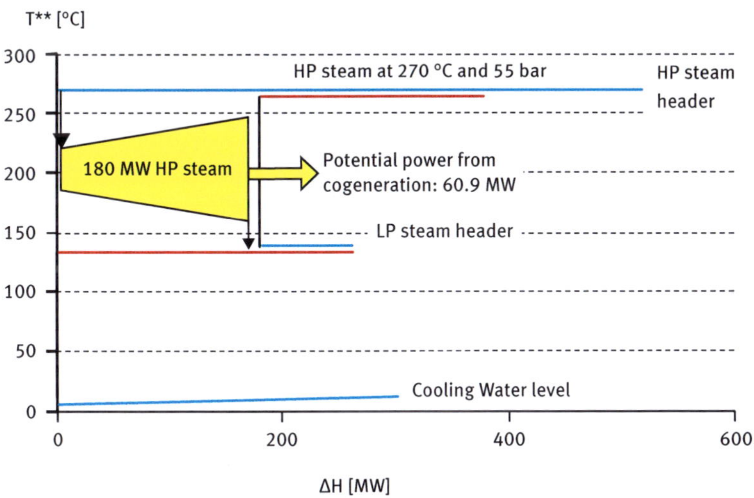

Fig. 4.19: The cogeneration target for Total Site Example 2.

Consider again the SUGCC of Example 2 from Fig. 4.18 It reflects a situation where the Heat Recovery through the steam system is maximised, corresponding to the minimum amount of HP steam that is required from the site boilers. In producing the Total Site Composite Curves, representing the situation of maximum Heat Recovery and minimal HP steam supply, the starting point was from the positions of the Total Site Profiles, which represents a situation of no Heat Recovery and maximum HP supply from the site boilers. Between these two extreme states there can be a variation of the HP steam supply from the boilers.

An even further increase of the HP steam supply from the boilers is possible. Based only on the consideration of the process steam demands and the Total Site Pinch such an increase makes no sense in terms of Heat Recovery and efficiency. However, if cogeneration is considered with total energy cost as a criterion, this involves cost items for power import from the grid or potential revenue from power export to the grid, in addition to the cost of fuel. Experience has shown (Smith and Varbanov, 2005) that there are certain price ratios at which such an increased steam generation, accompanied by on-site power generation by condensing steam turbines may be profitable. Such a potential situation for Total Site Example 2 would be reflected by a SUGCC like the one in Fig. 4.20, where about 60 MW extra steam is generated by the HP boiler and run through a condensing steam turbine. For more details on exploring such options, refer to the article on advanced Total Site Analysis by Varbanov et al. (2004b).

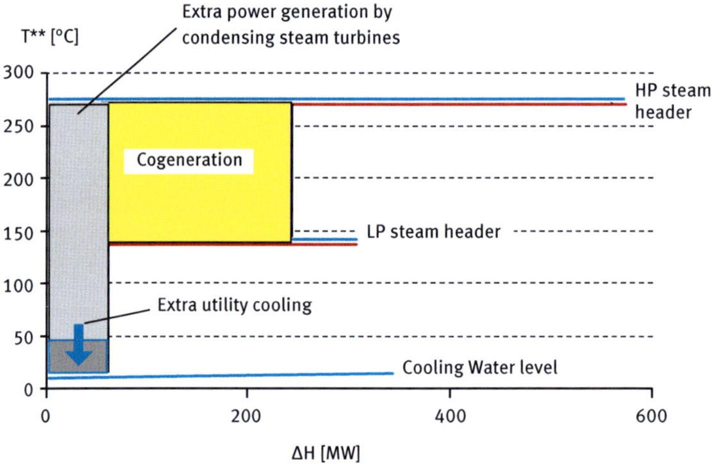

Fig. 4.20: Additional power cogeneration by condensing steam turbines for Total Site Example 2.

4.6.3 Choice of optimal steam pressure levels

Until this point, the steam pressure levels have been considered as fixed. However, broadening the view a little offers new opportunities at the stage of new system design and use of Total Site targeting. If the pressure levels of the intermediate utilities are allowed to vary, it is possible to achieve better Heat Recovery through the utility system (Klemeš et al., 1997). The fundamental trends of the Total Site Profiles and the derived Site Composite Curves can be exploited for this purpose. In this case, the steam pressure level becomes a degree of freedom to be exploited in the optimisation. The main trade-off in a Total Site Utility System with a single intermediate steam header is formed as follows (Fig. 4.21):

(a) If the pressure (consequently also the temperature) level of the header is lowered, it generally increases the potential amount of utility generation from Site Heat

Sources, which would be reflected by a wider horizontal segment of the Site Utility Generation Composite Curve.

(b) On the other hand, the trend regarding the Site Heat Sinks is the opposite – higher pressure level means larger demand to be covered by the current utility: lower level – smaller amount.

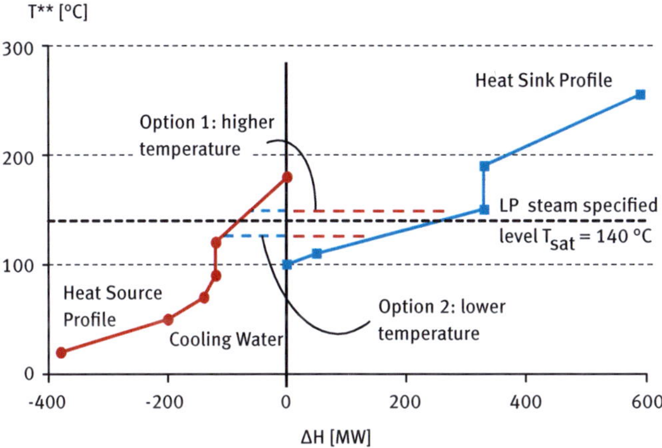

Fig. 4.21: Typical utility generation and use trends for Total Site Profiles in the case of Total Site Example 2.

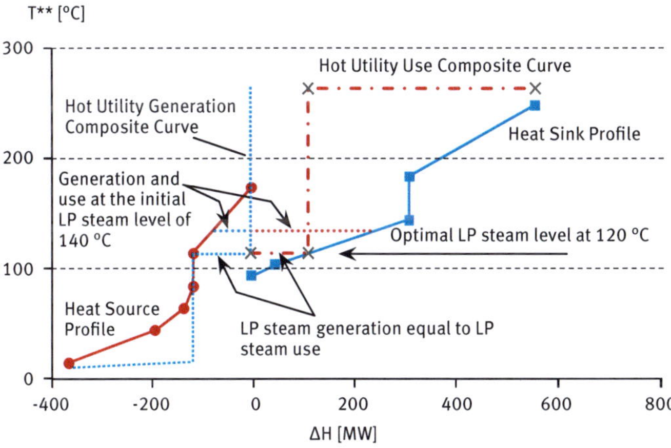

Fig. 4.22: Optimal saturation temperature of the LP steam header for Total Site Example 2.

Exploiting such a trade-off, it is possible to optimise the exact saturation temperature level and from there also the pressure of the particular utility. A defining property of this optimal level is that the steam generation by Site Heat Sources is equal to the steam use by Site Heat Sinks, which is logical – any extra generated utility also has to be used in order to effect Heat Recovery. Any unused generated utility would be anyway directed to the cooling water.

For the case of Total Site Example 2, the optimal LP steam saturation temperature with minimisation of the boiler steam use is at 120 °C as illustrated by the diagram in Fig. 4.22. The extent of the improvement is demonstrated in Table 4.18.

Tab. 4.18: Results of the LP steam level optimisation for Total Site Example 2 (all quantities are in MW).

	LP steam at 140 °C	LP steam at 120 °C	Change = new – old
LP steam generation	80	120	+40
LP steam use	260	120	−140
HP steam to processes	330	470	+140
HP steam target	**510**	**470**	**−40**
Cooling water target	**300**	**260**	**−40**

In the case when more than one intermediate utility is allowed on a site, the model is more complex and the optimisation would require using a mathematical optimisation solver. It is also important to mention that the level optimisation can be performed also accounting for cogeneration via the utility system, as has been described by Klemeš et al., (1997) and later elaborated by Shang and Kokossis (2004).

4.7 Advanced Total Site developments

4.7.1 Introduction of process-specific minimum allowed temperature differences

In the basic examples and considerations in this chapter, all processes on a site as well as all heat transfer operations between processes and utilities are assumed to have only one, single ΔT_{min} specification. Such an assumption may be too simplistic and lead to inadequate results due to imprecise estimation of the overall Total Site Heat Recovery targets.

When using a uniform ΔT_{min} specification, problems can arise in two cases:
- If the value of the common ΔT_{min} specification is overestimated for some parts of the Total Site Profiles, the achievable Heat Recovery targets for the process would be underestimated, resulting in higher energy cost estimates. This can have economic implications in the form of oversizing the utility system components – most notably the steam boilers, and lead to inefficient use of capital for the utility system.

- In the opposite case – if ΔT_{min} is set smaller than the optimal for the corresponding heat transfer type, the Heat Recovery targets would be overestimated leading to strive for more Heat Recovery site-wide and excessive capital costs for heat transfer area. In addition, the underestimated hot and cold utility demands may result in undersized utility facilities such as boilers and cooling towers. This, in turn, can potentially lead to either infeasible operational situations or at the very least reduced reliability of the utility system by having to operate near or at its maximum capacity.

This issue has been investigated by Varbanov et al. (2012) where a minimum allowed temperature difference specification is defined for each significant Heat Exchange case – inside each site process, plus the cases of generating or using each site utility on the interfaces with each process. The article provides a thorough formulation of the necessary additional Total Site concepts and a modified targeting procedure to suit the new degrees of freedom. These have been illustrated with a case study showing that the difference in the Total Site targets between the cases of single and multiple ΔT_{min} specifications can be as large as 30 % – see Fig. 4.23.

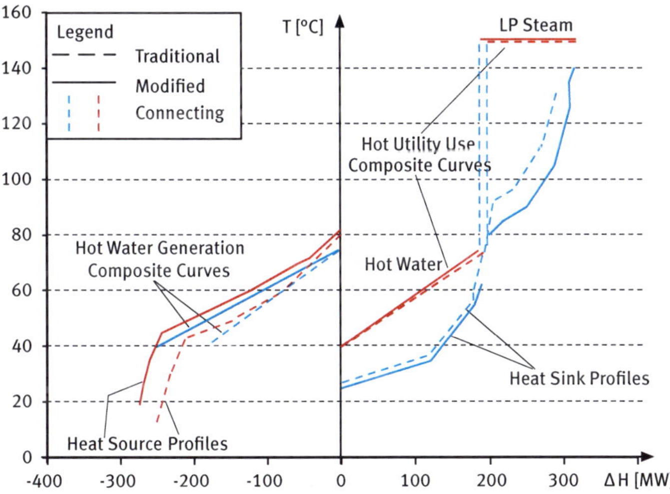

Fig. 4.23: Difference in the Total Site targets when allowing multiple ΔT_{min} specifications.

4.7.2 Numerical tools for Total Site Heat Integration

The Total Site targeting concept and tools discussed to this point bear the distinguishing elements of a powerful and elegant methodology for evaluating and optimising energy efficient processes. These include

- Thermodynamic soundness of the concept and the procedure
- Efficient visualisation tools – the Total Site Profiles and the Total Site Composite Curves
- Simplicity of construction and especially interpretation

These properties have brought about the increased popularity and applications of Process Integration. There have even been some reasonably successful software tools implementing the methodology – e.g. STAR (STAR, 2013) and its newer counterpart Site-int (Site-int, 2013). However, even these computer tools, which automate the construction of the Total Site Profiles and Site Composite Curves, their central concepts for computation, construction and display are the curves themselves. This results in both computational and results interpretation problems, preventing efficient further use of the Total Site concept.

An innovative step toward unlocking this hidden potential is a recent work by Liew et al. (2012). The study presents several new numeric based contributions:

1. Total Site Sensitivity Table (TSST), a tool for exploring the effects of plant shutdown or production changes on heat distribution and utility generation systems over a Total Site;
2. A new numerical tool for Total Site Heat Integration – the Total Site Problem Table Algorithm (TS-PTA), which extends the well-established Problem Table Algorithm (PTA) to Total Site Analysis;
3. A simple new method for calculating multiple utility levels in both the PTA and TS-PTA;
4. The Total Site Utility Distribution (TSUD) table, which can be used to design a Total Site Utility Distribution Network.

The key novelty from the enumerated contributions is the introduction of the Total Site Problem Table and the associated heat cascades, all supported by the numerical evaluation of the site targets for utilities. The introduction of this concept enabled sensitivity analysis for Total Sites as well as a follow-up work on utility system planning (Liew et al., 2013) – in terms of operation planning as well as development planning.

4.7.3 Power Integration

Pinch Analysis is a well-established methodology of Process Integration for designing optimal networks for recovery and conservation of resources such as heat, mass, water, carbon, gas, properties and solid materials for more than four decades. However its application to power (electricity) systems analysis is still under development.

Recent works of Wan Alwi et al. (2012) extend the Pinch Analysis concept used in Process Integration to determine the of minimum power (electricity) targets for systems comprising hybrid renewable energy sources. Power Pinch Analysis (PoPA) tools described in that paper include graphical techniques to determine the minimum target for outsourced power and the amount of excess power for storage during start up and normal operations. The PoPA tools can be used by energy managers, electrical and power engineers and decision makers involved in the design of hybrid power system. Graphical and numerical Power Pinch Analysis (PoPA) tools are systematic and simple to implement in the optimisation of power systems.

4.7.3.1 Methodology

This section describes the methodology for Power Pinch Analysis (PoPA) which consists of two main steps:
1. Data extraction, and
2. Targeting the *minimum outsourced electricity supply (MOES)* and the *available excess electricity for next day (AEEND)*.

Step 1: Data extraction

The first step of a Pinch Analysis application for resource conservation typically involves data extraction of a system's sources (resource availability) and demands (resource requirements). In the case of hybrid power systems, power sources are the instantaneous on-site electricity generation from the available renewable energy sources such as solar photovoltaic, wind or biomass. The power demands represent equipment electricity consumption that can be determined from equipment power ratings. The power sources and power demands are recorded at the time they are available. Tables 4.19 and 4.20 show the power sources and demands for the illustrative Case Study 1 that represents the case of an off-grid hybrid power system. In this case, excess power will be stored in a battery system. Power generated and consumed are calculated using Equations (4.11) and (4.12). Initially, the demands consumed 268 kW of outsourced electricity supply daily. The maximum power demand of 19 kW occurs between 8 to 18 h (see Table 4.21).

$$Power\ generation\ (kWh) = Power\ generated\ (kW) \times Time\ interval\ (h) \tag{4.11}$$

$$Power\ consumed\ (kWh) = Power\ rating\ (kW) \times Time\ interval\ (h). \tag{4.12}$$

To better illustrate the PoPA method using the illustrative case study, the average value of the generated renewable energy is considered for the specified time range. For real cases, RE supply fluctuates at each time interval and the designer should use the average amount of RE generated hourly to determine the minimum power targets using PoPA method.

Tab. 4.19: Power sources for Illustrative Case Study 1.

Power source	Time, h		Time interval, h	Power rating generated, kW	Power generation, kWh
	From	To			
Solar	8	18	10	5	50
Wind	2	10	8	5	40
Biomass	0	24	24	7	168

Tab. 4.20: Power demands for Illustrative Case Study 1.

Power demand appliances	Time, h		Time interval, h	Power rating, kW	Power consumption, kWh
	From	To			
Appliance 1	0	24	24	3	72
Appliance 2	8	18	10	5	50
Appliance 3	0	24	24	2	48
Appliance 4	8	18	10	5	50
Appliance 5	8	20	12	4	48

Step 2: Targeting the Minimum Outsourced Electricity Supply and Available Excess Electricity for the Next Day

This section introduces a new technique known as the Power Composite Curves (PCC) to determine the maximum power transfer from power sources to power demands, the MOES and AEEND for storage. The PCC is a graphical approach that is based on Pinch Analysis Composite Curves (Linnhoff et al., 1982, latest edition 1994).

Power Composite Curves

The PCC can be constructed as follows:

1. Y-axis represents the time scale from '0' to '24' h while x-axis represents the power generation or consumption.
2. An individual power source or power demand line is plotted as shown in Fig. 4.24. Note that the gradient of the line can be computed from the inverse of power rating for the corresponding power demand or power source.
3. The Composite Power Source (or power demand) line is obtained from the sum of power sources (or power demands) within a given time interval. The same procedure is repeated for the rest of the system time intervals in order to yield the Source and Demand Composite Curves for the entire system.

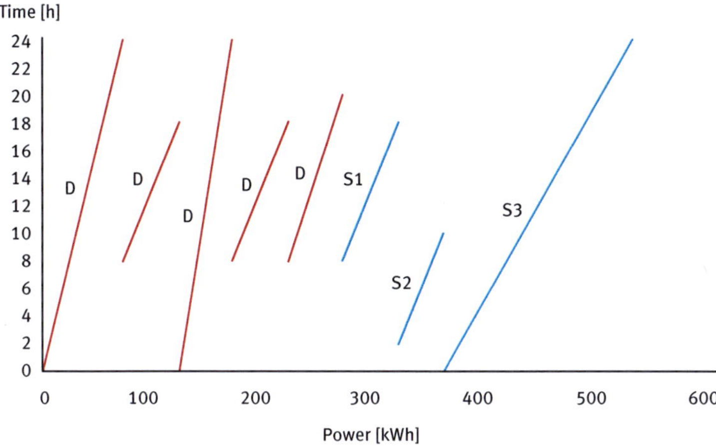

Fig. 4.24: Individual power source and demand lines for Illustrative Case Study 1.

4. The Pinch Point can be determined by shifting the Source Composite Curve to the right-hand side until it touches the Demand Composite Curve. Note that the Source Composite Curve has to be on the right-hand side of Demand Composite Curve since power can only be transferred from a current to a later time interval. A complete PCC is shown in Fig. 4.25.

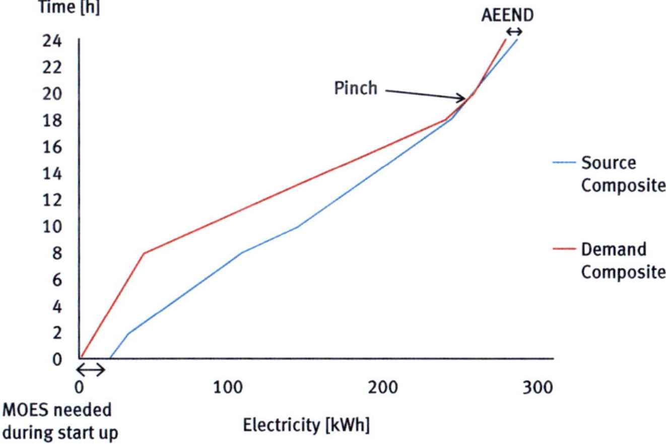

Fig. 4.25: Power Composite Curves (PCC) for a 24 h operation (Illustrative Case Study 1).

5. The excess demand line below the pinch point gives the MOES needed to be purchased during a system start up, and the excess power source above the Pinch

gives the AEEND. MOES and AEEND targets are 18 kW and 8 kW in illustrative Case Study 1.

6. Now it is possible to determine the amount of MOES needed during daily operation. This can be done by merging the PCC from Day 1 with the next day's PCC by linking the end of the SCC from the current-day PCC to the beginning of the source composite of the next day's PCC. Merging the current-day PCC with the next day's PCC gives the Continuous PCC (CPCC) – see Fig. 4.26. There are two possible scenarios for daily operations assumed:

(a) Scenario 1 is when AEEND (18 kW) is less than or equal to MOES (8 kW) for Day 1. It indicates that the amount of excess power from the previous day can be used to reduce the amount of outsourced electricity needed for the next day. Illustrative Case Study 1 is classified as Scenario 1. The excess demand line from the second PCC onwards will give the daily outsourced electricity supply needed to be purchased during normal operations. The CPCC reduces the MOES to 10 kW. The hybrid power system results in a savings of 96.3 % from MOES cost as compared to a conventional power system without RE.

(b) Scenario 2 is a case where AEEND is more than MOES. It indicates that the amount of excess power sources from the previous day is larger than the amount of outsourced electricity needed for the next day. For Scenario 2, if AEEND is continuously cascaded to the next day, it will cause accumulation of power inside the storage system. Scenario 2 is explained using Illustrative Case Study 2. In this case, the wind power source rating between times 2 to 10 h in illustrative Case Study 1 is increased from 5 kW to 6.5 kW. Fig. 4.27 shows the CPCC for Scenario 2. The excess source line after integration with power demand curve for the next day represents the excess wasted power source.

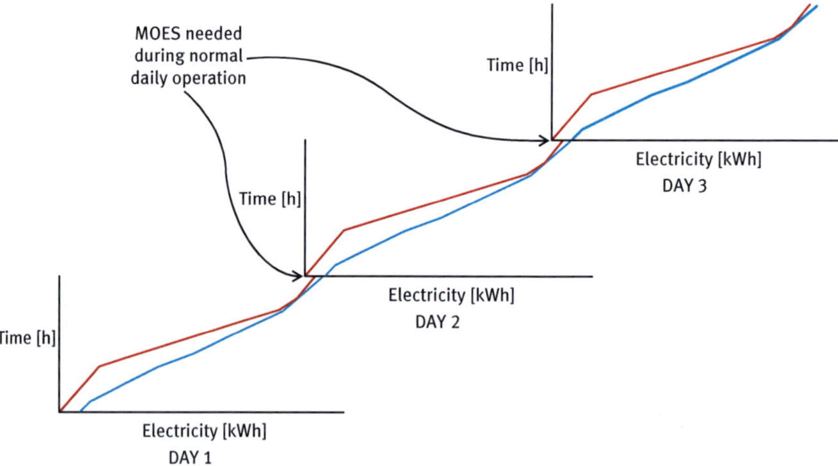

Fig. 4.26: Continuous Power Composite Curves (CPCC) for Illustrative Case Study 1.

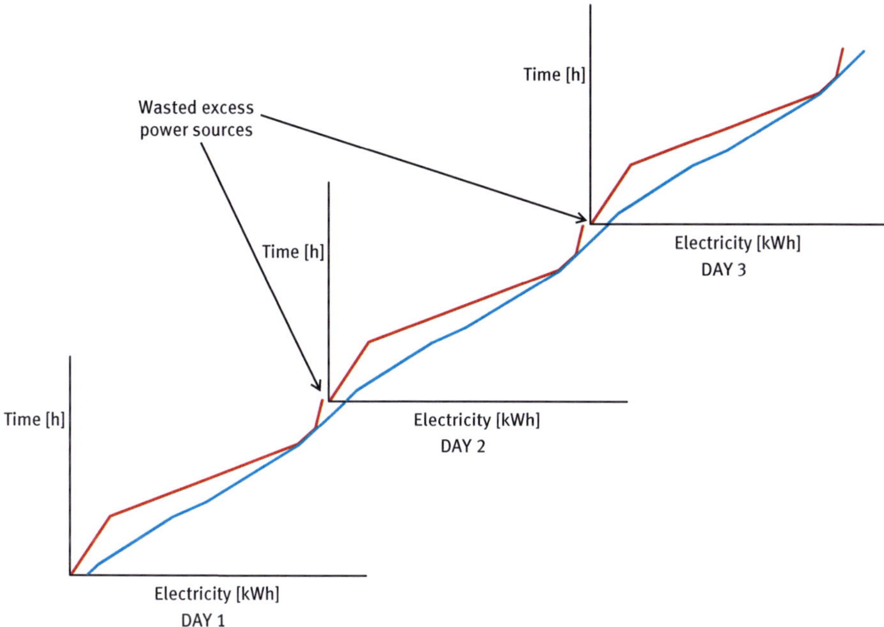

Fig. 4.27: Continuous Power Composite Curves (CPCC) for Illustrative Case Study 2.

However, the power losses incurred in the systems have been considered in the following work: Mohammad Rozali et al. (2013). This contribution extends the PoPA method by considering the power losses that occur in the power system conversion, transfer and storage. The losses' effect on the minimum outsourced electrical energy targets and storage capacity are evaluated.

This is a very recent part of the Total Site methodology and more information can be found in the Sources of Further Information Part.

4.8 Summary

This chapter provides a guide on the fundamentals of Total Site Integration. It discusses the techniques available for analysing Total Sites from the point of view of process Heat Sources and Heat Sinks, whose energy demands can be met to a large extent by Heat Recovery through the generation and use of intermediate utilities (mainly steam), thereby minimising the amount of steam that has to be supplied by central site utility boilers.

The extraction of Heat Sources and Sinks, after maximising process level Heat Recovery, can be accomplished with the aid of the individual process Grand Composite Curves and the results of the Problem Table Algorithm. The sources of heat from the individual site processes can be integrated to produce the Source Site Profile. Similarly the sinks of heat from the individual site processes can be integrated to produce the Sink Site Profile. Plotted together they make up the Total Site Profiles showing the total amount of heat available from all the processes on the site in relation to temperature, and the amount of heat needed by all the site processes.

The heat sources can be related to the generation of steam from the individual processes at various selected pressures, and the heat sinks can be related to the use of steam at various pressures. The use and generation of steam at these predetermined pressures in relation to the heat sources and sinks can be shown in the Total Site Profiles. Without any Heat Recovery between the generation and use of steam the Total Site Profiles can show the amount of steam that has to be supplied by central site utility boilers. However, the generation of steam from the process heat sources can be used to supply the process heat sinks, and consequently reduce the steam supply from the site utility boilers. The maximum amount of Heat Recovery from the site processes through the steam distribution system can be shown in the Total Site Composite Curves.

The generation and use of steam related to the process Heat Sources and Heat Sinks, and accompanied by the supply of steam from the site utility boilers, can be shown on the Site Utility Grand Composite Curve. This plot can also indicate the power generation potential that can accompany the steam generation and use at the various pressure levels.

References

Dhole, V.R. and Linnhoff, B. (1993). Total Site targets for fuel, co-generation, emissions, and cooling, *Computers and Chemical Engineering*, 17(Supplement), S101–S109.

Klemeš, J., Dhole, V.R., Raissi, K., Perry, S.J. and Puigjaner, L. (1997). Targeting and design methodology for reduction of fuel, power and CO_2 on Total Sites, *Applied Thermal Engineering*, 7, 993–1003.

Klemeš, J., Friedler, F., Bulatov, I. and Varbanov, P. (2010). *Sustainability in the Process Industry: Integration and Optimization*, New York, USA: McGraw Hill Companies Inc.

Linnhoff, B., Townsend, D.W., Boland, D., Hewitt, G.F., Thomas, B.E.A., Guy, A.R. and Marsland, R.H. (1982). *A User Guide to Process Integration for the Efficient Use of Energy*, Rugby, UK: IChemE, latest edition 1994.

Liew, P.Y., Wan Alwi, S.R., Varbanov, P.S., Manan, Z.A. and Klemeš, J.J. (2012). A numerical technique for Total Site sensitivity analysis, *Applied Thermal Engineering*, 40, 397–408.

Liew, P.Y., Wan Alwi, S.R., Varbanov, P.S., Manan, Z.A. and Klemeš, J.J. (2013). Centralised utility system planning for a Total Site Heat Integration network, *Computers & Chemical Engineering*, DOI: 10.1016/j.compchemeng.2013.02.007.

Matsuda, K., Hirochi, Y., Tatsumi, H. and Shire, T. (2009). Applying heat integration total site based pinch technology to a large industrial area in Japan to further improve performance of highly efficient process plants, *Energy*, 34(10), 1687–1692.

Mavromatis, S.P. and Kokossis., A.C. (1998). Conceptual optimisation of utility networks for operational variations – I. Targets and level optimisation, *Chem. Eng. Sci.*, 53(8): 1585–1608.

Mohammad Rozali, N.E., Wan Alwi, S.R., Manan, Z.A., Klemeš, J.J. and Hassan, M.Y. (2013). Process integration of hybrid power systems with energy losses considerations, *Energy*, 55, 38–45.

Raissi, K. (1994). *Total Site Integration*, PhD Thesis, UMIST, Manchester, UK.

Shang, Z. (2000). *Analysis and Optimisation of Total Site Utility Systems*, Ph Thesis, UMIST, Manchester, UK.

Shang, Z. and Kokossis, A. (2004). A transhipment model for the optimisation of steam levels of total site utility system for multiperiod operation, *Computers and Chemical Engineering*, 28(9), 1673–1688.

Site-int 2013. (2013). SITE-int. Process Integration Ltd. www.processint.com/chemical-industrial-software/site-int, accessed 31/11/2013.

Smith, R. and Varbanov, P. (2005). What's the price of steam? *Chemical Engineering Progress*, 101(7), 29–33.

STAR (2013). Process Integration Software (Centre for Process Integration, School of Chemical Engineering and Analytical Science, University of Manchester, UK). www.ceas.manchester. ac.uk/media/eps/schoolofchemicalengineeringandanalyticalscience/content/researchall/ centres/processintegration/STAR.pdf (accessed 21/09/2013).

Varbanov, P.S., Doyle, S. and Smith, R. (2004a), Modelling and optimisation of utility systems, *Trans IChemE, Chem Eng Res Des*, 82(A5), 561–578.

Varbanov, P., Perry, S., Makwana,Y., Zhu, X.X. and Smith, R. (2004b), Top-level analysis of site utility systems, *Trans IChemE, Chem Eng Res Des*, 82(A6), 784–795.

Varbanov, P., Perry, S., Klemeš, J. and Smith, R. (2005). Synthesis of industrial utility systems: Cost-effective de-carbonization, *Applied Thermal Engineering*, 25(7), 985–1001.

Varbanov, P.S., Fodor, Z, Klemeš, J.J (2012). Total Site targeting with process specific minimum temperature difference (ΔT_{min}), *Energy*, 44(1), 20–28, doi:10.1016/j.energy.2011.12.025.

Wan Alwi, S.R., Mohammad Rozali, N.E., Manan, Z.A. and Klemeš, J.J. (2012). A process integration targeting method for hybrid power systems, *Energy*, 44(1), 6–10.

5 Introduction to Water Pinch Analysis

5.1 Water management and minimisation

The industrial sector is one of the major consumers of water. According to UNESCO (2003), industry consumes 22 % of world water, while 70 % is consumed by the agriculture sector and 8 % by the domestic sector. Water is used for cooling and heating, for cleaning, for product formulation, as a transportation as well as a mass separating agent. Rising water prices and wastewater treatment costs, stricter environmental regulations for wastewater quality discharge and shortage of clean water resources have spurred efforts by industryto employ effective means to reduce water footprint via water management and minimisation strategies that include:

1. Improving the efficiency of water usage
2. Reducing water wastage
3. Applying water-saving technologies
4. Reuse, recycling and regeneration (treatment) of wastewater
5. Outsourcing of available water sources e.g. rainwater, river, snow or sea water.

The terminology 'water footprint' was introduced by Hoekstra (2003). The water footprint of a product is the volume of freshwater used to produce the product, measured over the full supply chain after considering the direct and indirect water uses. There are three types of water footprints:

1. Blue water footprint – Consumption of blue water resources (surface and groundwater) along the supply chain of a product.
2. Green water footprint – Consumption of green water resources (rainwater before it becomes run-off water).
3. Grey water footprint – Volume of freshwater that is required to absorb the load of pollutants given the natural background concentrations and existing ambient water quality standards.

Table 5.1 shows the global average water footprints for several products.

Tab. 5.1: Global average water footprints for several products (Water Footprint, 2012).

Product	Global Average Water Footprint, L/kg	Green	Blue	Grey
Butter	5,553	85 %	8 %	7 %
Bioethanol (from sugar cane)	2,107	66 %	27 %	6 %
Chocolate	17,196	98 %	1 %	1 %
Rice	2,497	68 %	20	11 %
Sugar (from sugar cane)	1,782	66 %	27 %	6 %

5.2 History and definition of Water Pinch Analysis

Water reuse and recycling involves using wastewater as a water source for a process or an activity that does not require freshwater. This is possible if the mass load of the wastewater source is equal to or less than the mass load required by the water sink (i.e., the wastewater source is of the same quality, or cleaner than the water sink). Water can also be reused and recycled by mixing wastewater of different concentrations with freshwater to obtain the desired mass load as well as flowrate of the water sink.

Industry typically consists of many processes that use water and produce wastewater. This creates the need for a systematic method to analyse all the water streams and to design the maximum water recovery network that utilises the minimum external freshwater and generates the minimum wastewater. Water Pinch Analysis (WPA) was developed in 1994 by Wang and Smith (1994a) to achieve the aforementioned goal.

Initially adapted from the concept of mass exchange network synthesis that was established by El-Halwagi and Manousiouthakis (1989). The earlier emphasis of the WPA was on mass transfer-based water-using processes where water is mainly used as a mass separating agent (a lean stream) to remove impurities from a rich stream – e.g. in solvent extraction, gas absorption and vessel cleaning, as shown in Fig. 5.1. This case is also widely referred to as the fixed load (water minimisation) problem. The main goal of WPA at that time was to target the minimum freshwater required to remove a fixed amount of impurities. This method assumes that there are no water losses or gains, and that the amount of freshwater consumed is equal to the amount of wastewater generated.

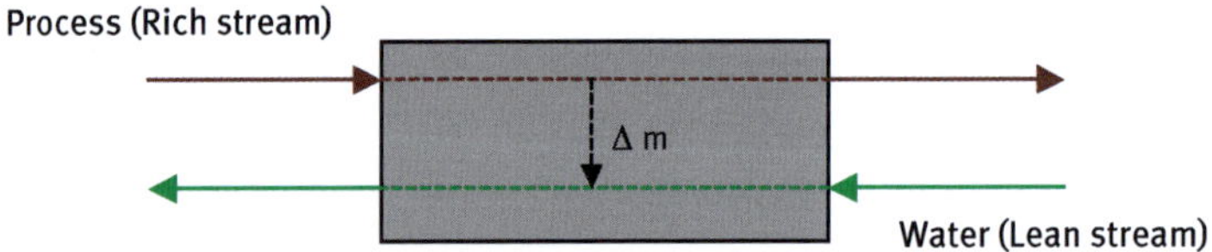

Fig. 5.1: Transfer of species from a rich to lean stream in a mass exchanger.

Later, Dhole et al. (1996) extended this method to a wider application that includes non-mass transfer-based processes. This case is widely known as the fixed flowrate (water minimisation) problem. Non-mass transfer-based processes include water consumed or generated from a reaction, water losses from evaporation etc. In this new version of WPA, water sources and sinks are considered as separate streams during data extraction (as shown in Fig. 5.2).

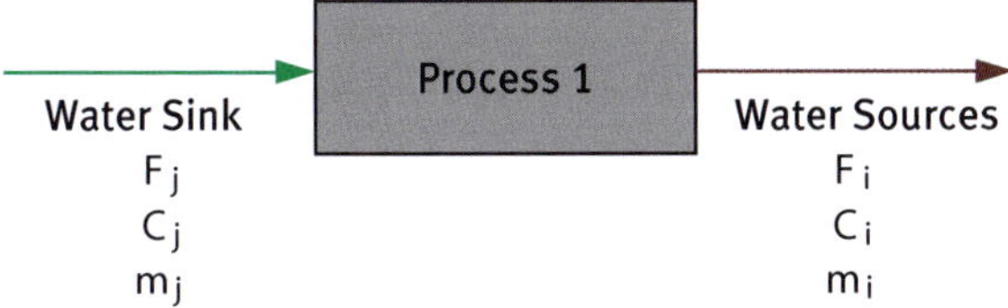

Fig. 5.2: Water source and demand.

WPA can be defined as a systematic technique of implementing water minimisation strategies through integration of processes to achieve the maximum water efficiency (Manan et al., 2004). The term "Pinch" refers to the limit or "bottlenecks" for water recovery. By using WPA, a designer can quickly determine the maximum amount of water recovery possible within a water system. WPA provides a target or "benchmark" for industry to achieve a "step change", as opposed to incremental water reduction by designing an optimal water recovery network that implement various water recovery strategies. In a water minimisation project, industry personnel will typically ask the following questions:

- *What could you possibly do (that we have not done)?*
- *How much water could be further saved?*
- *How high are the monetary savings?*
- *How much investment is needed?*
- *Is it going to be worthwhile?*
- *How could the savings be realised?*

By using WPA, all these questions can be answered. The following chapters highlight how WPA answers all these questions.

5.3 Applications of Water Pinch Analysis

WPA has been proven to give beneficial water savings. The potential savings from WPA application have been widely reported. Table 5.2 lists some of the industrial WPA applications reported in the literature.

Table 5.2 shows that the range of savings achieved for the reference cases are between 7 % to 72 % freshwater reduction. The amount of savings depends on the extent of water reuse in the industry as well as on the quantity and quality of water sources and sinks available. The payback periods reported were within two years, which is indeed attractive for most industries. WPA is highly recommended for water-intensive industries especially in countries where the water tariffs and wastewater treatment costs are high.

Tab. 5.2: Applications of Water Pinch Analysis.

Company	Process/Industry	Location	Flow Reduction, %	Reference
Confidential	Chemical & Fibres	Germany	25	Tainsh and Rudman (1999)
Cerestar	Corn Processing	UK	25	
Gulf Oil	Oil Refining	UK	30	
Monsanto	Chemicals	UK	40	
Parenco	Paper Mill	Netherlands	20	
Sasol	Coal Chemicals	South Africa	50	
Unilever	Polymers (batch)	UK	60	
US Air Force	Military Base	USA	40	
Confidential	Oil Refining	Netherlands	40	
Confidential	Chemicals	USA	40	
Confidential	Chemicals & Fibres	USA	25	
Confidential	Semiconductors	Malaysia	72	Wan Alwi and Manan (2008)
Confidential	Chloralkali	Malaysia	7	Handani et al. (2010)
Confidential	Paper Mill	Malaysia	13	Manan et al. (2007)
Confidential	Beet Sugar	Slovenia	69	Zbontar Zverand and Glavič (2005)
Confidential	Citrus	Argentina	30	Thevendiraraj et al. (2003)

5.4 Water Pinch Analysis steps

WPA involves five key steps as follows:

Step 1: Analysis of water network
The existing or the base case water network is analysed during a plant audit. All the water-using operations are identified and water balances are developed. The scope of analysis is also defined.

Step 2: Data extraction
Water sources and water sinks having potential for reuse and recycling are identified. The limiting water flowrate and limiting concentration data are extracted.

Step 3: Setting the minimum utility targets
The minimum freshwater requirement and wastewater generation, or the maximum water recovery targets are established using either the graphical or numerical targeting techniques. This will be further elaborated in Chapter 6.

Step 4: Water network design/retrofit
Design a water recovery network to achieve the minimum water targets. This will be further elaborated in Chapter 7.

Step 5: Economics and technical evaluations.
Water Pinch Analysis (WPA) has been applied for grassroots and retrofit cases. In the case of retrofit, repiping cost may be needed. Retrofitting may also cause process disruption. All these factors must be considered before undertaking a WPA project. The final step of WPA involves evaluating the economics and the technical constraints of the water network. Mathematical modelling techniques have been widely used to calculate the network cost by considering the repiping needs, the geographical layout and constraints, and the associated fixed as well as variable costs. For preliminary economic calculations, payback period is widely used as a criterion to assess the feasibility of a proposed network solution. The payback period is calculated using Equation (5.1). The net capital investment is the investment needed for pipe rerouting, pumping, and storage tanks for intermittent processes and control systems. The net annual savings are the potential savings from freshwater reduction.

$$Payback\ period\ (yrs) = \frac{Net\ Capital\ Investment\ t(\$)}{NetAnnualSavings(\$/y)} \tag{5.1}$$

Other than payback period, net present value (NPV) as well as return on investment (ROI) are also widely used. In order to evaluate the technical feasibility, the amount of water saving, the technology risk, the implementation period and the impact on normal operation need to be evaluated.

5.5 Analysis of water networks and data extraction

Industrial implementation of WPA begins with the analysis of water network and water stream data extraction. These steps, which are the critical path of a WPA project, are described in detail next.

5.5.1 Analysis of water networks

The first step in the implementation of WPA involves analysing the existing water network of a facility by developing an overall water mass balance. This includes identifying where all the freshwater is being distributed to the processes, and from which processes the wastewater is being generated. Process Flow Diagrams (PFDs), Process Instrumentation Diagrams (P&IDs) or the DCS system will help, but in most cases not all water-using processes are recorded in these diagrams. Therefore, line tracing still needs to be performed.

Water is used for various uses in industry such as for cooling and heating, transportation agent, cleaning, product formulation, mass separating agent, general plant services, and potable/sanitary. There are two important parameters which need to be determined for WPA, i.e. flowrate and contaminant concentration of each of the water streams.

5.5.1.1 Water flowrate determination

Extracting water flowrate data is often not so straightforward because the data for water inlets and outlets are not normally monitored or recorded in a plant. Freshwater is taken from the water mains and thereafter, distributed throughout a plant. There are typically no devices installed to measure the flowrate of the distributed water sources. Wastewater from various processes is typically collected in a common drain, and the quantity of the mixed wastewater is only known at the wastewater treatment plant. The mixed wastewater is no longer useful for WPA implementation because the quality of this wastewater is at its lowest as it has already been degraded as a result of being mixed with all sorts of spent water. For WPA implementation, the flowrate of segregated individual wastewater streams generated from each process or utility stream needs to be extracted.

As a start, water balance data can be obtained from the existing process material balances, computer monitoring, routine measurement, previous plant studies and laboratory reports (Liu et al., 2004). If the water data are still unavailable, a plant owner can decide either to install a fixed flow meter or use a portable ultrasonic flow meter to determine the missing flowrates of the water sources and sinks. However, the limitation of using an ultrasonic flow meter is that the pipe coating and insulation may need to be scrapped (this is not preferred by most plant owners). In addition, the readings may not be accurate if a pipe is not full, or when there is the possibility of fouling in the pipe. Knowledgeable estimates can be made as a last resort. For water data that fluctuates according to the users' pattern of water usage, e.g. for potable water usages such as toilet flushing, water surveys can also be conducted to estimate the average or the range of water consumption.

Stream data should be extracted during normal operations, and not during plant shutdowns, start-ups or upsets. Water balance should be conducted to account for all water flows. As a general rule, the aim is to complete at least 80 % of the site water balance. The priority is to select streams that have large flowrates and/or contribute significantly to the contaminant loading. Streams have been verified to have relatively much smaller flowrates can be ignored.

5.5.1.2 Contaminant concentration determination

Contaminant concentration data are also not easily available in a plant. For water sinks (or water streams entering a process unit), the contaminant concentration limit can be extracted from equipment specification sheets, engineers' estimate or literature review. Other sources of constraints include (Foo et al., 2006):

1. The manufacturer's design data;
2. Physical limitations such as weeping flowrate, flooding flowrate, channelling flowrate, saturation composition, minimum mass transfer driving force, maximum solubility;
3. Technical constraints such as to avoid scaling, precipitation, corrosion, explosion, contaminant build up etc.

As an example, according to a maintenance manual, the water quality standard for a cooling tower should be maintained at a pH between 7.2 to 8.5, total hardness less than 200 ppm, total alkalinity less than 100 ppm, chloride ion less than 200 ppm, total ion (Fe) less than 1.0 ppm, silica less than 50 ppm, ammonium ion less than 1.0 ppm, and total dissolved solids (TDS) less than 1500 ppm. These data can be taken as the contaminant limits for the cooling tower water sink.

For water sources (or wastewater leaving a process), water samples can be taken from a sampling point and the contaminant concentration can be determined by a lab test or by using a water-quality test kit. Several water samples should be taken from each water source in order to determine the reliable upper and lower bounds of the contaminant concentration. The maximum outlet concentration for water sources may also need to be adjusted to take into account the mass pick up. For example, if contaminants are dissolving, then an increase in the inlet concentration will also be reflected at the outlet concentration to balance the system.

5.5.2 Data extraction

In a water-intensive process plant, specifying the limiting data can be a very tricky, arduous and time-consuming task. This is typically the bottleneck, and more importantly, the critical success factor for a water minimisation project. Once the water balances have been developed, the next step is to select the potential candidates for water reuse, i.e. to extract the limiting water data. Processes chosen are preferably located close together and chemically related. This can potentially reduce the piping as well as pumping costs, and allow the use of a common water regeneration unit. Almost similar chemical properties will also facilitate water integration.

The system can be modelled as a single contaminant or multiple contaminant problem, based on the water quality requirement of a process plant. However, the problems involving multiple contaminants necessitates a complex modelling procedure and has to rely on mathematical modelling and optimisation techniques. Hence, the typical simplifying assumption is the use of aggregated contaminants, such as suspended solids and total dissolved solids (TDS), that allow the consideration of multiple quality factors to be modelled as a single contaminants system (The Institution of Chemical Engineers, 2000). The aggregated contaminants modelled as a single contaminant system are known as a pseudo-single contaminant system.

The primary contaminants that prevent direct reuse in the water system are then chosen. As an example, for a semiconductor plant, conductivity is the most important contaminant which is monitored throughout the system and is selected for WPA, implementation even though there are also other contaminants in the water system. The proposed network however still needs to be reassessed by checking that all other contaminant concentrations not considered are still within allowable limits before implementing WPA.

The water sources' and sinks' flowrates and quality requirements for each of the selected water-using processes are then extracted. Most processes normally operate between the upper and lower bounds of its process parameters to maintain the product quality while avoiding the equipment from overloading. For operability reasons, the maximum concentration limit and the minimum flowrate limit are extracted to ensure that the worst case scenario is considered when designing the water reuse network.

Once the limiting water flowrate and contaminant concentration have been extracted, the contaminant mass load of the stream can be calculated. The mass load for each water-using process can be calculated by using Equation (5.2).

$$m = \frac{F \times C}{1000}. \tag{5.2}$$

Where, m = contaminant mass load, kg/h
F = flowrate, t/h
C = contaminant concentration, ppm
C_{in} = inlet contaminant concentration, ppm
C_{out} = outlet contaminant concentration, ppm

5.5.3 Example

5.5.3.1 Example 5.1 – Speciality chemical plant

Figs. 5.3(a) and 5.3(b) show the utility and process flowsheets of speciality chemical plant case study taken from Wang and Smith (1995). Suspended solids were chosen as the main contaminant of the system. The original water balance data extracted from the flowsheet is shown in Table 5.3.

Initially all the existing inlet concentrations are zero, meaning that freshwater is used throughout all existing systems (see Table 5.3). However, this constraint needs to be re-evaluated as it does not allow for any water reuse. The maximum inlet concentrations were re-evaluated to include the appropriate reuse constraints such as

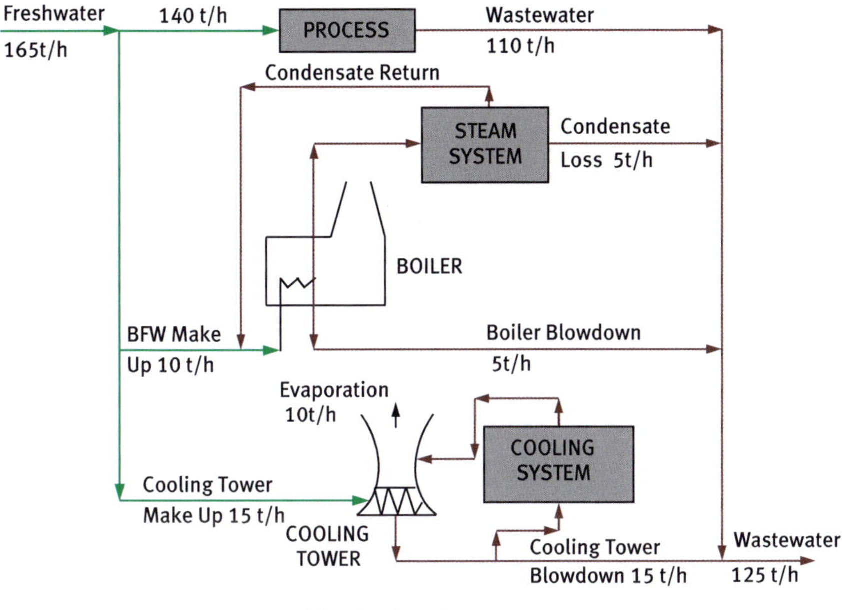

(a) Utility flowsheet

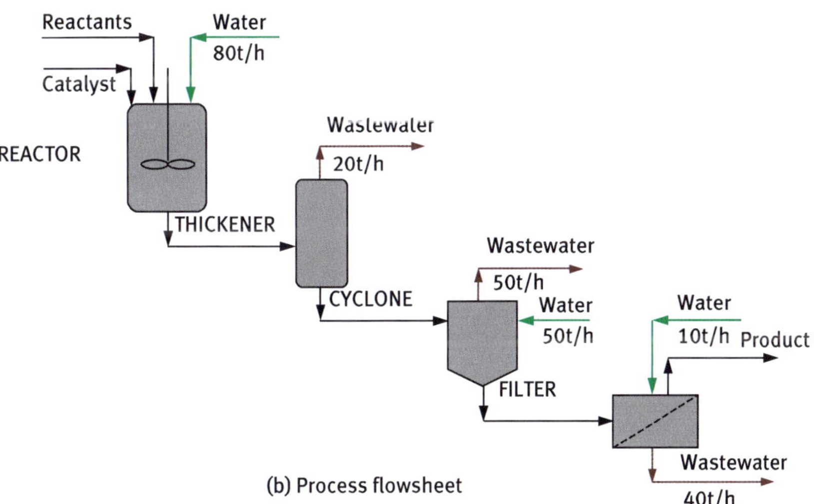

(b) Process flowsheet

Fig. 5.3: Example 5.1. Speciality chemical plant.

those based on solubility limits, corrosion limits, and contamination of products. The maximum outlet concentrations were also adjusted to take into account the mass pick up due to the increase in the maximum inlet concentration. For example, if contaminants are dissolving, then an increase in the inlet concentration will also be reflected at the outlet concentration to balance the system. Table 5.4 shows the adjusted maximum inlet and outlet concentrations.

Tab. 5.3: Water balance for Example 5.1.

Operation	F_{in} (t/h)	F_{out} (t/h)	C_{in} (ppm)	C_{out} (ppm)
Reactor	80	20	0	900
Cyclone	50	50	0	500
Filtration	10	40	0	100
Steam System	10	10	0	10
Cooling System	15	5	0	90

Tab. 5.4: Modified stream data for Example 5.1.

Operation	F_{in} (t/h)	F_{out} (t/h)	C_{in} (ppm)	C_{out} (ppm)
Reactor	80	20	100	1,000
Cyclone	50	50	200	700
Filtration	10	40	0	100
Steam System	10	10	0	10
Cooling System	15	5	10	1,00

Note that water inlet and outlet are not the same for the reactor (water loss), filtration (water gain) and cooling system (water loss). The outlet and inlet flowrates are extracted as the source and sink data, as shown in Table 5.5. The mass load can be calculated by using Equation (5.2) as shown in the last column of Table 5.5.

Tab. 5.5: Source and Sink data extraction.

Operation	Sink	F (t/h)	C (ppm)	m (t/h)	Source	F (t/h)	C (ppm)	m (t/h)
Reactor	SK1	80	100	8	SR1	20	1,000	20
Cyclone	SK2	50	200	10	SR2	50	700	35
Filtration	SK3	10	0	0	SR3	40	100	4
Steam System	SK4	10	0	0	SR4	10	10	0.1
Cooling System	SK5	15	10	0.15	SR5	5	100	0.5

5.5.3.2 Example 5.2 – Acrylonitrile (AN) production plant

Fig. 5.4 shows an Acrylonitrile (AN) production plant flowsheet from El-Halwagi (1997), with its water balances. Ammonia (NH_3) was identified as the main contaminant in the water system that limits water reuse. The plant wished to expand its AN production. However, their biotreatment facility was already operating at full capacity. The plant had to either reduce their wastewater discharge or build an additional wastewater treatment plant. The plant decided to implement WPA. In order to reduce wastewater discharge and avoid building a new treatment plant.

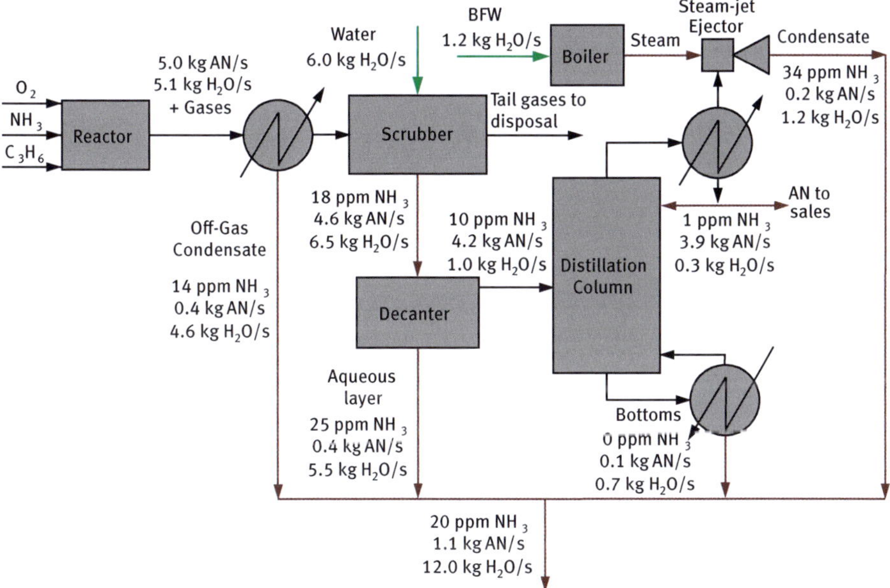

Fig. 5.4: Flowsheet of AN production for Example 5.2.

From the flowsheet, two water sinks can be observed. The first is the scrubber, which uses freshwater to remove the reactor remaining off-gas. The second sink is the boiler feed water (BFW) which is used in the boiler to generate steam for the steam jet ejector. Four streams are identified as water sources. These include wastewater generated from the off-gas condensate, aqueous layer of the decanter, bottom product of the distillation column and condensate from the steam ejector. Note that data for the individual wastewater streams were extracted as opposed to the mixed stream going to the wastewater treatment plant.

Note that water that is further used in subsequent processes, for example between the decanter and the distillation column, is not extracted as a water source or sink as it does not require freshwater and does not generate wastewater. Water coming from the distillation top product, which is part of the product to be sold, is also not considered a water source as it is not sent to the wastewater treatment plant.

Below are the technical constraints for the water sinks.

1. Scrubber
   - $5.8 \leq$ flowrate of wash feed (kg/s) ≤ 6.2
   - $0.0 \leq NH_3$ content of wash feed (ppm) ≤ 10.0
2. Boiler feed water
   - NH_3 content = 0.0 ppm
   - AN content = 0.0 ppm

For the scrubber, an upper and lower bound was given for both the flowrate and contaminant concentration. As explained in the previous section, the minimum flowrate and maximum contaminant concentration are extracted as the limiting water data, i.e. 5.8 kg/s and 10 ppm. Tables 5.6 and 5.7 show the extracted limiting water sink and source data for the AN plant. The mass load of each water source and water sink is calculated by using Equation (5.2) and shown in the last column.

Tab. 5.6: Limiting water sinks data for AN production plant (Example 5.2).

Sink	Description	F, kg/s	C, ppm	m, mg/s
SK1	Scrubber	5.8	10	58
SK2	Boiler	1.2	0	0

Tab. 5.7: Limiting water sources data for AN production plant (Example 5.2).

Source	Description	F, kg/s	C, ppm	m, mg/s
SR1	Distillation bottoms	0.8	0	0
SR2	Off-gas condensate	5.0	14	70.0
SR3	Aqueous layer	5.9	25	147.5
SR4	Ejector condensate	1.4	34	47.6

References

Dhole, V.R., Ramchandani, N., Tainsh, R.A. and Wasilewski, M. (1996). Make your process water pay for itself, *Chemical Engineering*, 103, 100–103.

El-Halwagi, M.M. and Manousiouthakis, V. (1989). Synthesis of mass-exchange networks, *AIChE Journal*, 35 (8), 1233–1244. DOI: 10.1002/aic.690350802.

El-Halwagi, M.M. (1997). *Pollution Prevention Through Process Integration: Systematic Design Tools*, San Diego, California, USA: Academic Press.

Foo, D.C.Y. (2009). State-of-the-art review of Pinch Analysis techniques for water network synthesis, *Industrial and Engineering Chemistry Research*, 48 (11), 5125–5159.

Handani, Z.B., Wan Alwi, S.R., Hashim, H. and Manan, Z.A. (2010). A holistic approach for design of minimum water networks using Mixed Integer Linear Programming (MILP) Technique, *Ind. Eng. Chem. Res.* 49, 5742–5751.

Hoekstra, A.Y. (ed.) (2003). Virtual water trade: Proceedings of the International Expert Meeting on Virtual Water Trade, Value of Water Research Report Series No 12, UNESCO-IHE, Delft, Netherlands, www.waterfootprint.org/Reports/Report12.pdf, Accessed 1/1/2013.

Liu, Y.A., Lucas, B. and Mann, J. (2004). Up-to-date tools for water-system optimization, *Chemical Engineering Magazine*, January 2004, 30–41.

Manan, Z.A., Tan, Y.L., Foo, D.C.Y.and Tea, S.Y. (2007). Application of water cascade analysis technique for water minimisation in a paper mill plant, *International Journal of Pollution Prevention*, 29 (1–3), 90–103.

Manan, Z.A., Tan, Y.L. and Foo, D.C.Y. (2004). Targeting the minimum water flowrate using water cascade analysis technique, *AIChE Journal*, 50 (12), 3169–3183. DOI: 10.1002/aic.10235.

Noureldin, M.B.and El-Halwagi, M.M. (1999). Interval-based targeting for pollution prevention via mass integration, *Computers and Chemical Engineering*, 23, 1527–1543.

Tainsh, R.A. and Rudman A.R. (1999). Practical techniques and methods to develop an efficient water management strategy, Linnhoff March International, Paper presented at: IQPC conference *"Water Recycling and Effluent Re-Use"*, in: KBC Advanced Technologies PLC. www.dev-kbcat. contentactive.com/default/documents/technical papers/IQPC_Water_Paper.pdf, Accessed on 22 March 2012.

The Institution of Chemical Engineers (2000). Guide to Industrial Water Conservation. Part 1. Water Re-use, in: *Single Contaminant Systems*, ESDU 0020, Rugby, UK.

Thevendiraraj, S., Klemeš, J., Paz, D., Aso, G. and Cardenas, J. (2003). Water and Wastewater Minimisation Study of a Citrus Plant, *Resources, Conservation and Recycling*, 37, 227–250.

UNESCO (2003). "Water for People, Water for Life", United Nations World Water Development Report, www.unesdoc.unesco.org/images/0012/001295/129556e.pdf, Accessed on 9 Dec 2012.

Wan Alwi, S.R. and Manan, Z.A. (2008). A holistic framework for design of cost-effective minimum water utilisation network. *Journal of Environmental Management*, 88, 219–252. DOI: 10.1016/j. jenvman.2007.02.011.

Wang, Y.P. and Smith, R. (1994b). Design of distributed effluent treatment systems. *Chem. Eng. Sci.*, 49, 3127–3145. DOI: 10.1016/0009–2509(94)E0126-B.

Wang, Y.P. and Smith, R. (1994a). Wastewater minimisation. *Chemical Engineering Science*, 49(7), 981–1006. DOI: 10.1016/0009–2509(94)80006–5.

Wang, Y.P. and Smith, R. (1995). Wastewater minimization with flowrate constraints. *Chem. Eng. Res. Des.*, 24, 2093–2113.

Water Footprint, Product Gallery. www.waterfootprint.org/?page=files/productgallery, Accessed on 9 December 2012.

Zbontar Zver, L. and Glavic, P. (2005). Water minimization in process industries: Case study in beet sugar plant. *Resources, Conservation and Recycling*, 43 (2), 133–145.

6 Setting the maximum water recovery targets

6.1 Introduction

Upon completion of the limiting water data extraction as explained in Chapter 5, the maximum water recovery (MWR) targets can now be determined.

Techniques for MWR targeting include:

1. Limiting Composite Curves (Fig. 6.1) by Wang and Smith (1994) – Water-using processes are plotted on a concentration versus flowrate diagram. This method is applicable for both fixed flowrate and fixed load problems. However, for fixed flowrate problems, it involves the pairing of sources and sinks to include the water losses and water gains.

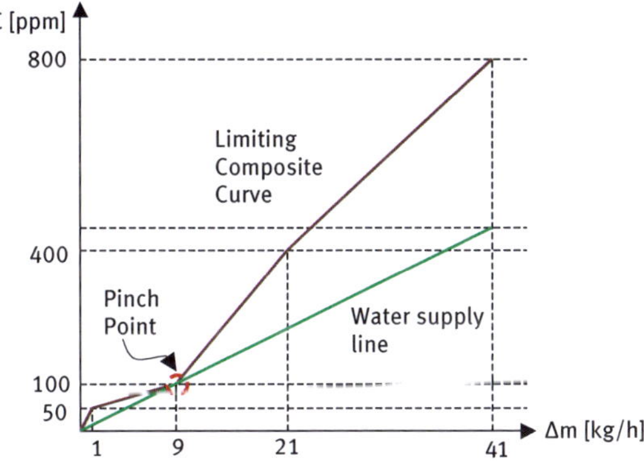

Fig. 6.1: Limiting Composite Curve.

2. Water Surplus Diagram (Fig. 6.2) by Hallale (2002) – Water sources and sinks are plotted separately and plotted on a concentration versus flowrate diagram to determine the water surplus and deficit. The water surplus and deficit are then summed to form the Water Surplus Diagram (WSD). The freshwater targets are then predicted by using a trial and error method until all the WSD lie on the right-hand side of the y-axis. This method is applicable for both fixed flowrate and fixed load problems.

3. Source/Sink Composite Curves (Fig. 6.3) by El-Halwagi et al. (2003) and Prakash and Shenoy (2005a) – Water sources and sinks are plotted on a contaminant mass load versus flowrate. It was first introduced by El-Halwagi et al. (2003) for material recycle/reuse. However, later Prakash and Shenoy (2005a) introduced an almost similar method for water systems. The Source/Sink Composite Curves overcome the limitation of WSD which requires two graphs and an iterative process. This method is applicable for both fixed flowrate and fixed load problems.

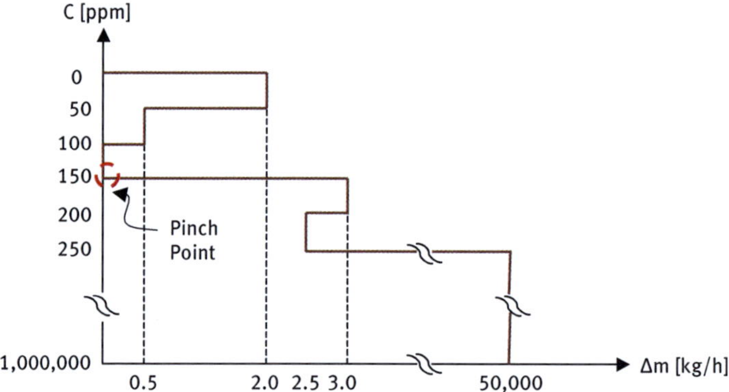

Fig. 6.2: Water Surplus Diagram.

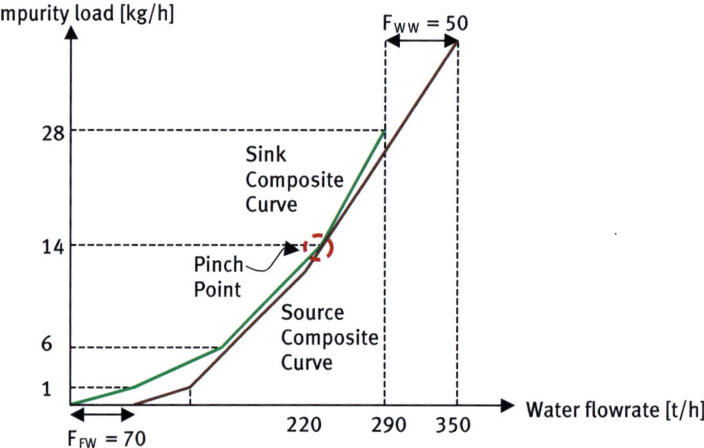

Fig. 6.3: Source/Sink Composite Curve.

4. Water Cascade Analysis (WCA) (Fig. 6.4) by Manan et al. (2004) and Foo et al. (2006) – The original WCA that was based on purity intervals (Manan et al., 2004) was a numerical version of WSD by Hallale (2002). WCA eliminates the iterative process of WSD and provides more accurate results. Foo et al. (2006) later used concentration intervals in the WCA technique. Since the WCA is based on algebraic calculations, its steps and formulas can be readily programmed into Microsoft Excel. This allows the tabulated values of WCA to be easily duplicated to other rows by using the Excel formula drag function.

k	C_k	$\sum_j F_j$	$\sum_i F_i$	$\sum_i F_i - \sum_j F_j$	$F_{C,k}$	Δm_k	Cum Δm_k	$F_{FW,k}$
k	C_k	$[\sum_j F_j]_k$	$[\sum_i F_i]_k$	$[\sum_i F_i - \sum_j F_j]_k$	F_{FW}	Δm_k		
$k+1$	C_{k+1}	$[\sum_j F_j]_k$	$[\sum_i F_i]_{k+1}$	$[\sum_i F_i - \sum_j F_j]_{k+1}$	$F_{C,k}$	Δm_{k+1}	Cum Δm_k	$F_{FW,k}$
					$F_{C,k+1}$		Cum Δm_{k+1}	$F_{FW,k+1}$
$\vdots$	$\vdots$	$\vdots$	$\vdots$	$\vdots$	$\vdots$	$\vdots$		$\vdots$
$n-1$	C_{n-1}	$[\sum_j F_j]_{n-1}$	$[\sum_i F_i]_{n-1}$	$[\sum_i F_i - \sum_j F_j]_{n-1}$		Δm_{n-1}	Cum Δm_n	$F_{FW,n}$
n	C_n	$[\sum_j F_j]_n$	$[\sum_i F_i]_n$	$[\sum_i F_i - \sum_j F_j]_n$	$F_{C,n-1} = F_{WW}$			

Fig. 6.4: Water Cascade Table.

5. Algebraic Targeting Approach (Fig. 6.5) by Al-Mutlaq et al. (2005) – A numerical version of Source/Sink Composite Curves. It uses the load interval diagram.
6. Source Composite Curves (Fig. 6.6) by Bandyopadhyay (2006) – A hybrid between numerical and graphical approaches. It uses an almost similar cascading approach as WCA but only requires single cascading instead of double cascading. The results of the numerical step is then used to plot Source Composite Curves, which is a plot of concentration versus mass load. The plot consists of a Source Composite Curve and a wastewater line. The advantage of this method is that it can predict the average outlet wastewater concentration.

k	C_k	$\sum_i F_i - \sum_j F_j$	Cum $F_{C,k}$	Δm_k	Cum Δm_k	$F_{WW,k}$
k	C_k	$[\sum_i F_i - \sum_j F_j]_k$	$[\sum_i F_i - \sum_j F_j]_k$	Δm_k		
$k+1$	C_{k+1}	$[\sum_i F_i - \sum_j F_j]_{k+1}$	Cum $[\sum_i F_i - \sum_j F_j]_{k+1}$	Δm_{k+1}	Cum Δm_{k+1}	$F_{WW,k+1}$
$\vdots$	$\vdots$	$\vdots$	$\vdots$	$\vdots$	Cum Δm_{n-1}	$F_{WW,n-1}$
$n-1$	C_{n-1}	$[\sum_i F_i - \sum_j F_j]_{n-1}$	Cum $[\sum_i F_i - \sum_j F_j]_{n-1}$	Δm_{n-1}	Cum Δm_n	$F_{WW,n}$
n	C_n	$[\sum_i F_i - \sum_j F_j]_n$				

Fig. 6.5: Algebraic steps for Water Source Diagram.

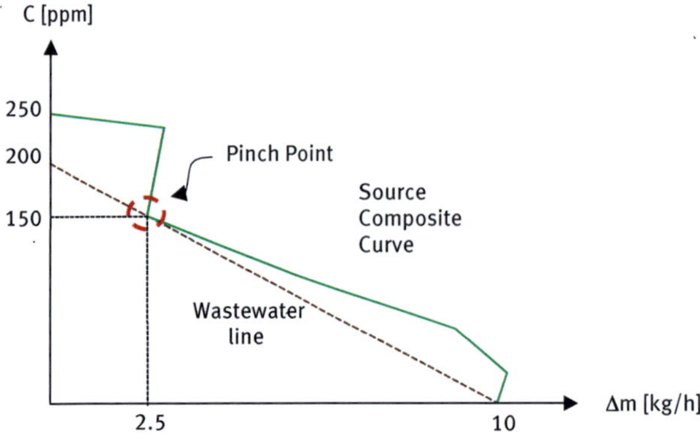

Fig. 6.6: Source Composite Curve.

All these MWR methods apply to continuous processes. For batch processes, these methods can also be used by employing the use of storage tanks. Since water sources may be generated at different times compared to when water is needed at the water sinks, a storage tank can be used to store the water source until it is needed at the water sinks. The use of storage tanks allows WPA to be applied to continuous or batch processes. However, there are also methods to target the maximum water recovery for batch processes without the use of storage tanks (direct reuse). For more details on batch maximum water recovery targeting, readers can refer to the work of Foo et al. (2005) who presented a time-dependent Water Cascade Analysis (WCA) technique and Majozi et al. (2006) who presented a graphical technique.

Section 6.2 provides detailed descriptions of the graphical Source/Sink Composite Curves (SSCC) and the algebraic Water Cascade Analysis (WCA) technqniques that have been widely used to determine the maximum water recovery target for single pure freshwater as the available utility. The WCA technique is easier to construct and gives more precise and accurate results due to its numerical nature. Its steps and formulas can be readily programmed into Microsoft Excel. This allows the tabulated values from the WCA to be easily duplicated to other rows by using the Excel formula drag function. On the other hand, the SSCC provides useful visualisation insights for the engineers to understand the proposed solution in order for them to influence the design.

For simplification, in most WPA applications, freshwater is typically assumed as the only water utility available at zero contaminant concentration even though freshwater may contain some small amount of contaminants in practice. Furthermore, various freshwater sources such as demineralised, deionised and potable water may also be available as utilities. Apart from that, "outsourced water" – i.e. water from the environment, that includes rainwater, snow, borehole water, river water and even "imported" spent water, may also be available for use in the plant area. In Section 6.3, the use of SSCC to determine the water targets for cases involving impure freshwater as well as multiple water sources, are described.

6.2 Maximum water recovery target for single pure freshwater

6.2.1 Water Cascade Analysis technique

Water Cascade Analysis (WCA) (Manan et al., 2004) is an algebraic or numerical targeting method that is used to determine the minimum water targets, i.e. the overall freshwater requirement and wastewater generation for a process after looking at the possibility of using the available water sources within a process to satisfy the water sinks. To achieve this objective, the net water flowrate, water surplus and deficit at the different water concentration levels within the process under study have to be established. The WCA was initially developed by using purity levels as the water quality measure instead of concentrations. Foo et al. (2006) later simplified the WCA method by using the concentration levels.

Fig. 6.7 is a conceptual illustration of how water cascading can minimise freshwater needs and wastewater generation. In Fig. 6.7(a), 100 kg/s of wastewater is produced by Operation 1 water source at the concentration level of 100 ppm and 50 kg/s. Water is needed by Operation 2 water sink at the concentration level of 200 ppm. Without considering water reuse, 100 kg/s of wastewater would be generated while 50 kg/s of freshwater would be required. However, as shown in Fig. 6.7(b), by making use of 100 kg/s of the water source at the concentration level of 100 ppm from Operation 1 to satisfy the water sink of 50 kg/s at the concentration level of 200 ppm from Operation 2, it is possible to avoid sending part of the water source directly to the effluent. Doing so not only reduces the wastewater generation but also the freshwater consumption, in both cases, by 50 kg/s.

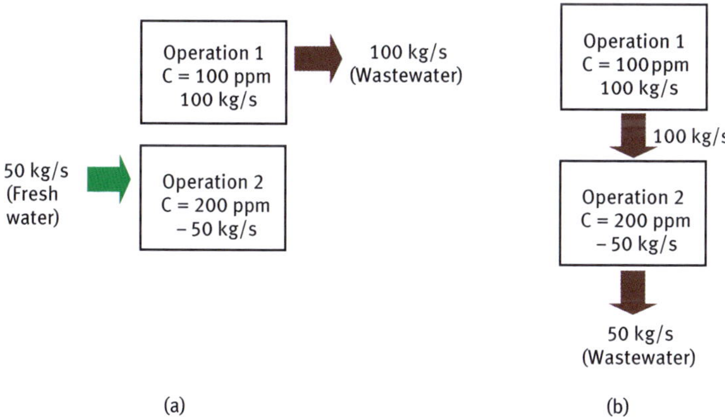

Fig. 6.7: The principle of water cascading.

Example 6.1 describes the construction of the Water Cascade Table (WCT) that is represented by Table 6.1.

Tab. 6.1: Limiting water data for Example 6.1 (Polley and Polley, 2000).

D_j	F_j, t/h	C, ppm	m, t/h
1	50	20	1
2	100	50	5
3	80	100	8
4	70	200	14

S_i	F_i, t/h	C, ppm	m, t/h
1	50	50	2.5
2	100	100	10
3	70	150	10.5
4	60	250	15

Tab. 6.2: Water cascade table for Example 6.1.

C_k, ppm	ΔC_k, ppm	ΣF_{SKi}, t/h	ΣF_{SRj}, t/h	$\Sigma F_{SKi} + \Sigma F_{SRj}$, t/h	F_C, t/h	Δm, kg/h	Cum. Δm, kg/h	$F_{FW, cum}$, t/h	F_C, t/h
									$F_{FW} = 70$
0				0			0		
	20				0	0			70
20		−50		−50			0	0	
	30				−50	−1,500			20
50		−100	50	−50			−1,500	−30	
	50				−100	−5,000			−30
100		−80	100	20			−6,500	−65	
	50				−80	−4,000			−10
150			70	70			−10,500	−70	(PINCH)
	50				−10	−500			60
200		−70		−70			−11,000	−55	
	50				−80	−4,000			−10
250			60	60			−15,000	−60	
					−20	0			$F_{WW} = 50$
							−15000		

To construct the WCT, the contaminant concentrations (C) of the water streams are listed in ascending order (see Column 1 of Table 6.2). Duplicate concentration should be listed only once. Column 2 lists the concentration difference (ΔC) computed using Equation (6.1).

$$\Delta C = Cn - C_n + 1.$$ (6.1)

The water sinks (F_{SK}) and sources (F_{SR}) flowrates in columns 3 and 4 are added at each concentration level in columns $\left(\sum_j F_{SK,j} \right)$ and $\left(\sum_i F_{SR,i} \right)$. For simplicity, $\left(\sum_i \right)$ be designated as Σ next. Note that water sinks are assigned with negative values while the water sources are positive. Water sinks and sources are summed in Column 5 ($\Sigma F_{SKi} + \Sigma F_{SRj}$) at each concentration level. A positive value in this column indicates a net surplus of water present at the respective concentration level, while a negative value indicates a net deficit of water. Any water sources at a lower concentration can be used as a source for water sinks at a higher concentration.

In Column 6, a zero freshwater flowrate is first assumed. This freshwater flowrate is then cascaded with Column 5 to give the cumulative flowrate (F_C) for each concentration level. The first row in this column represents the estimated flowrate of freshwater required for the water-using processes (F_{FW}). The total cumulative water flowrate value in the final column represents the total wastewater generated in the process (F_{WW}). However, this is the preliminary infeasible cascade and not the true freshwater and wastewater targets.

In Column 7, the product of the cumulative flowrate and concentration difference ($F_C \times \Delta C$) is calculated at every concentration level to give the mass load (Δm). The mass load is then cumulated down each concentration level ($Cum\ \Delta m$) in Column 8. An interval freshwater flowrate ($F_{FW,K}$) is then determined by using Eqn. (6.2) where C_{FW} is the freshwater concentration (Column 9). For this Example 6.1, C_{FW} of 0 ppm is assumed.

$$F_{FW,k} = \frac{cum\ \Delta m_k}{C_k - C_{FW}}$$ (6.2)

If negative value exists in this column, this means that there is insufficient water purity in the networks. Thus, more freshwater needs to be added until no negative value exists in this column. The largest negative value of $F_{FW,k}$ is taken and this value is replaced with the earlier assumed zero freshwater flowrate in Column 6 to be recascaded to obtain the feasible water cascade (Column 10). Note that the location of the concentration level with the largest negative value of $F_{FW,k}$ is also the location of the Pinch Point, i.e. at 150 ppm. The water source which exists at the Pinch is called the Pinch-causing source. Part of this source is located above the Pinch and part of it below the Pinch.

The new cascade now gives the minimum freshwater target (in the first row) and wastewater target (last row) of 70 t/h and 50 t/h.

6.2.2 Source/Sink Composite Curves (SSCC)

The SSCC is a plot of mass load (m) versus flowrate (F). It is used to target the minimum usage of fresh resources for material recycle/reuse networks (El-Halwagi, 2003). Example 6.1 is used to illustrate the steps to construct the SSCC. The Sink Composite Curve is created by connecting each sink with its corresponding mass load and flowrate cumulatively in ascending concentration order. The same is performed for the Source Composite Curve. This is shown in Fig. 6.8(a). The Source Composite Stream is then shifted to the right until it touches the Sink Composite Stream, with the Source Composite Curve located below the Sink Composite Curve in the overlapped region (see Fig. 6.8(b)). The Pinch occurs where the two Composite Curves touch. The Pinch concentration is determined by the concentration of the Pinch-causing source stream,

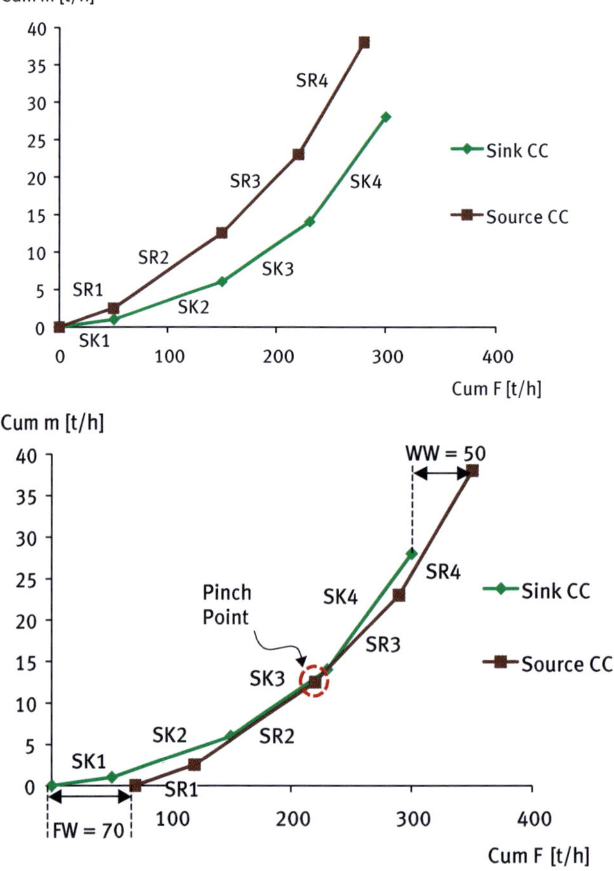

Fig. 6.8: Source/Sink Composite Curve for Example 6.4. (a) Before shifting Source Composite Curve, (b) After shifting Source Composite Curve.

i.e. S3 at 150 ppm. The minimum freshwater target is the flowrate distance difference between the beginning of the Source Composite Curve and the Sink Composite Curve. The minimum wastewater target is the flowrate distance difference between the end of the Source Composite Curve and the Sink Composite Curve. This is illustrated in Fig. 6.8(b). The freshwater and wastewater targets are the same as those obtained using the WCT, i.e. 70 t/h and 50 t/h.

6.2.3 Significance of the Pinch region

As in the case of Heat Pinch Analysis, the Pinch point in WPA is crucial in guiding designers towards the correct network design. Fig. 6.9 shows the definition of the Pinch region for a Multiple Pinch case.

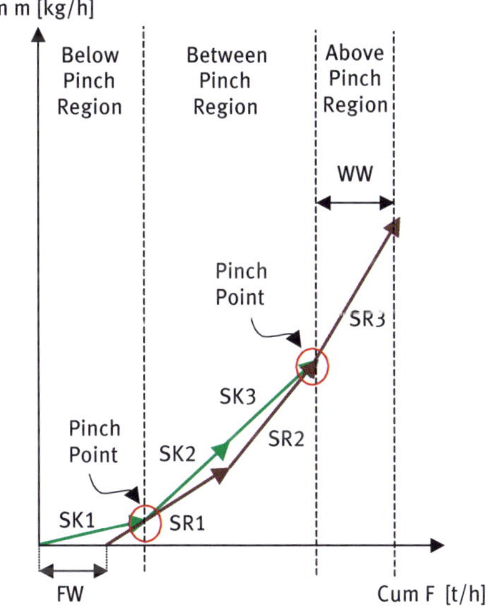

Fig. 6.9: Source/Sink Composite Curve – Pinch region classification.

A water source located on one side of the Pinch should only be used to satisfy a water sink located in the same Pinch region, or otherwise a water penalty will occur. An exception is the water source which is the Pinch-causing stream, as part of it is located above the Pinch while the other part is below the Pinch. Following is the significance of the Pinch regions:

1. Below the Pinch region (lower concentration level)

    All water sinks' mass load and flowrate should be satisfied either by water sources or freshwater. No wastewater should be generated.

2. Between Pinch region
   All the water sources should satisfy the water sinks' flowrate and mass load. No freshwater should be added or wastewater generated.
3. Above the Pinch region (higher concentration level)
   The water sources which satisfies the water sinks can have an equal or lower mass load amount. No freshwater should be added to this region. Excess water sources will be discharged as wastewater.

6.3 Maximum water recovery target for a single impure freshwater source

Wan Alwi and Manan (2007) proposed a method to determine the maximum water recovery target for a single impure freshwater source that may exist at a concentration lower or higher than other streams' concentrations. The authors divided the problem into Pinched and threshold problems as described next.

6.3.1 Pinched problems

Prior to determining the minimum new utilities flowrate (F_{MU}), it is important to establish if a utility is suitable for a given process, using the following heuristic:

Heuristic 6.1:

Only consider a water source as a utility if its concentration is lower than the concentration of the Pinch.

A utility at a concentration higher than the Pinch point will only increase wastewater. Given that a water source at a concentration lower than the Pinch concentration is available for the limiting data in Example 6.1 (Table 6.1). The F_{MU} can be obtained by systematically moving the source line above (SLA) along the utility line and the source line below (SLB) until Utilities/Process Pinches are obtained. Definitions of the source lines above (SLA) and below (SLB) the U are shown in Fig. 6.10.

Systematic shifting of SLA and SLB to get F_{MU} involve two key steps:
1. Move the Utility Line along with SLB to the right-hand side of the Sink Composite Curve until the line meets either the first Utility or Process Pinch (SLB and utility lines must Pinch the Sink Composite Curve).
2. Shift SLA upwards along the utility line until the entire SLA Pinches the Sink Composite Curve.

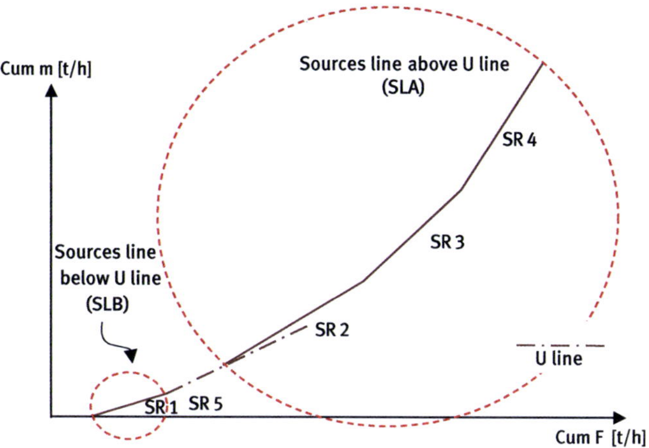

Fig. 6.10: Location of various water sources relative to the Utility Line, SR5.

1.   Example A: A utility with no SLB creates a Utility Pinch.

SR5 at a concentration of 10 ppm is a utility added to the limiting water data from Example 6.1. There is no water source below SR5. Hence, the SR5 line is drawn first and shifted until a Utility Pinch (C_{Pinch} = 10 ppm) occurs at cum m = 0 according to Step 1 (Fig. 6.11).

SLA (SR1 to SR4) is moved upwards according to Step 2 along the SR5 line until a process Pinch occurs at C_{Pinch} = 150 ppm (Fig. 6.11). The horizontal gap between the Sink Composite and the intersection of SLA and SR5 gives the minimum utility flow-rate (F_{MU}), i.e. 75 t/h.

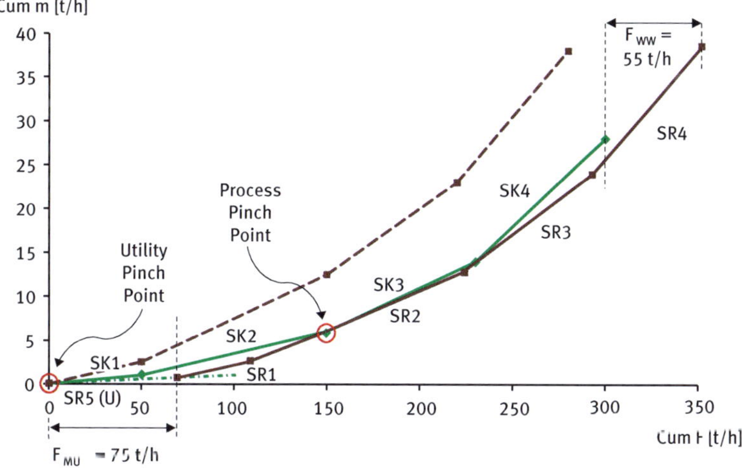

Fig. 6.11: SLA shifted along SR5. Final Composite Curve with minimum utility addition.

2. Example B: A utility with an SLB creates a utility Pinch.

SR5 was a utility at 80 ppm added to the limiting data from Polley and Polley (2000). Shifting SLB (S1) and SR5 along zero 'Cum m' axis according to Step 1 created a Utility Pinch at C_{Pinch} = 80 ppm (Fig. 6.12). SLA (SR2 to SR4) was next shifted according to Step 2 along SR5 from the new Pinch point onwards until a process Pinch point was created at C_{Pinch} = 100 ppm (Fig. 6.13). Fig. 6.14 shows the final Composite Curves with F_{MU} of 43.75 t/h.

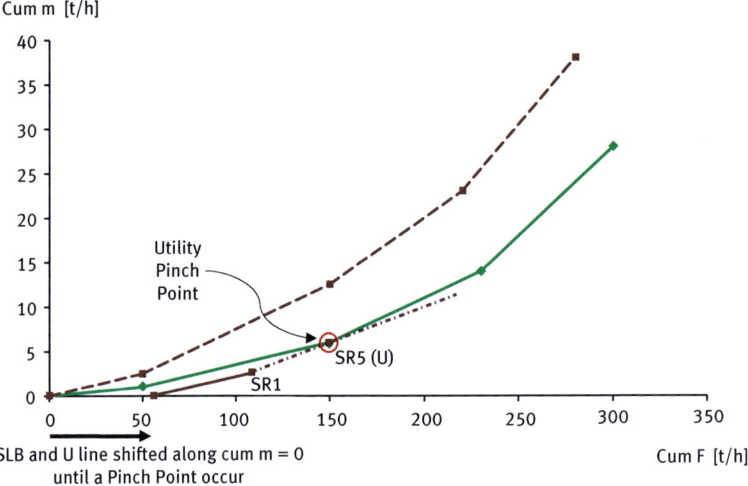

Fig. 6.12: SLB (S1) and SR5 shifted along the Cum m =0 line.

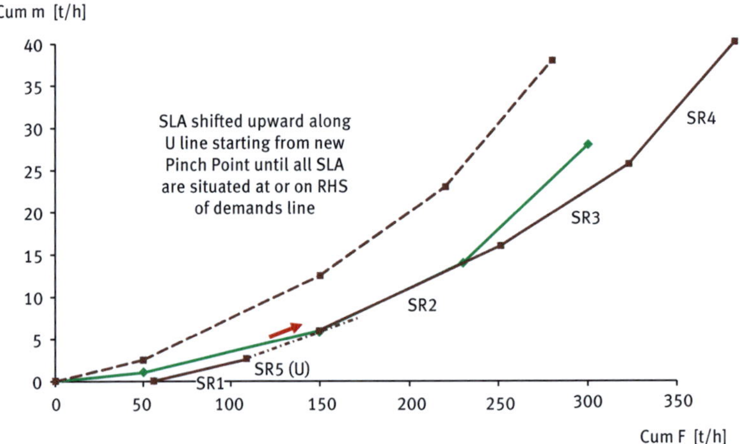

Fig. 6.13: SLA (S2 to S4) shifted upwards along SR5 from the new Pinch Point until SLA created another Pinch Point at C_{pinch} = 100 ppm.

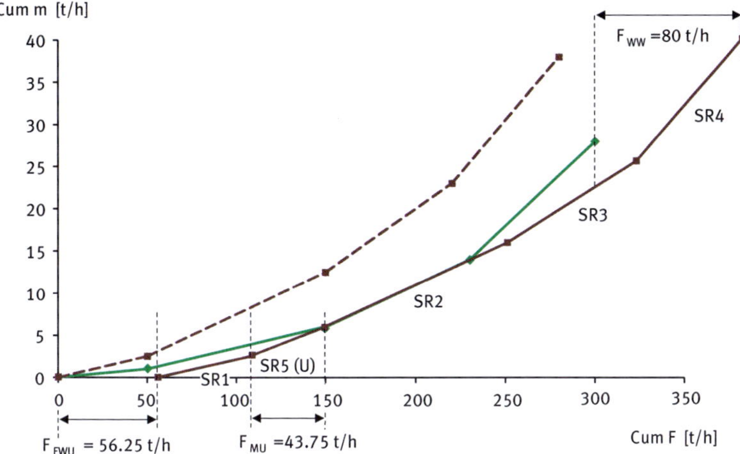

Fig. 6.14: Final Composite Curves with addition of S5.

3.   Example C: The utility is a regenerated water source that creates a Utility Pinch. The utility may be a water source regenerated from a concentration above to below the Pinch point. This may create new utility and process Pinch points at lower concentrations and reduce the length of the regenerated source line. Table 6.3 shows the limiting data from Sorin and Bedard (1999) which resulted in multiple Pinch points at 100 ppm and 180 ppm. To have beneficial water savings, a source above or at concentration of 180 ppm should be regenerated to a concentration below 100 ppm. SR6 is a new Utility created by regenerating SR5 from 250 to 30 ppm. SR6 is then shifted along

Tab. 6.3: Limiting data for Example 6.2 from Sorin and Bedard (1999).

Sink	F, kg/s	C, ppm	m, kg/s	Cum F, kg/s	Cum m, kg/s
SK1	120	0	0	120	0
SK2	80	50	4	200	4
SK3	80	50	4	280	8
SK4	140	140	19.6	420	27.6
SK5	80	170	13.6	500	41.2
SK6	195	240	46.8	695	88
Source					
SR1	120	100	12	120	12
SR2	80	140	11.2	200	23.2
SR3	140	180	25.2	340	48.4
SR4	80	230	18.4	420	66.8
SR5	195	250	48.75	615	115.55

the freshwater line until a Utility Pinch occurred at $C_{Pinch} = 30$ ppm (see Fig. 6.15). The SLA (SR1 to SR5) is next shifted along SR6 with SR5 original flowrate maintained until a Process Pinch occurs at 100 ppm (see Fig. 6.16). For $F_{MU} = 114.3$ kg/s, the amount of SR5 after reduction is calculated at 80.7 kg/s. Fig. 6.17 shows the final Composite Curves after SR5 reduction. The minimum freshwater and wastewater flowrates are 120 kg/s and 40 kg/s.

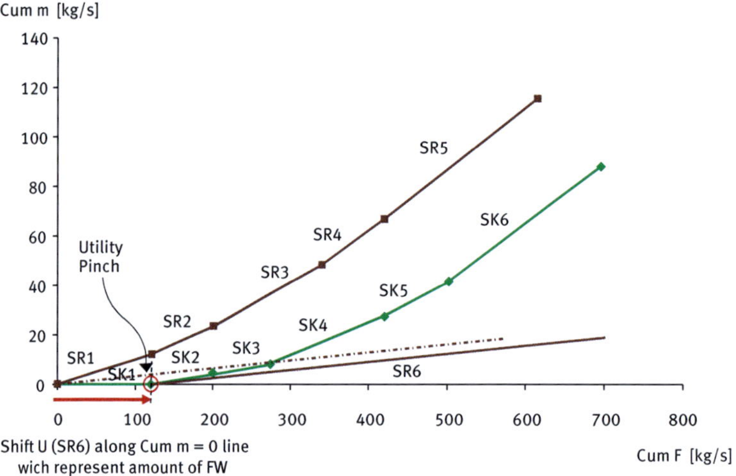

Fig. 6.15: SR6 utility line shifted along cum m =0 line.

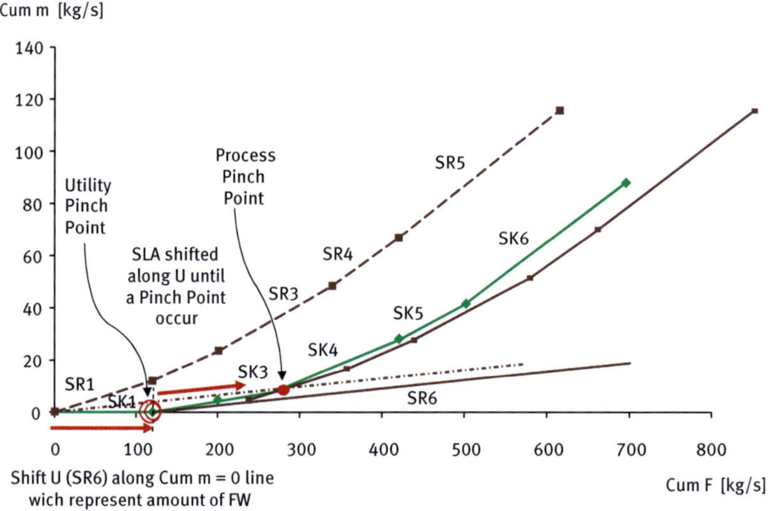

Fig. 6.16: SLA shifted along SR6 until a Pinch Point occurred.

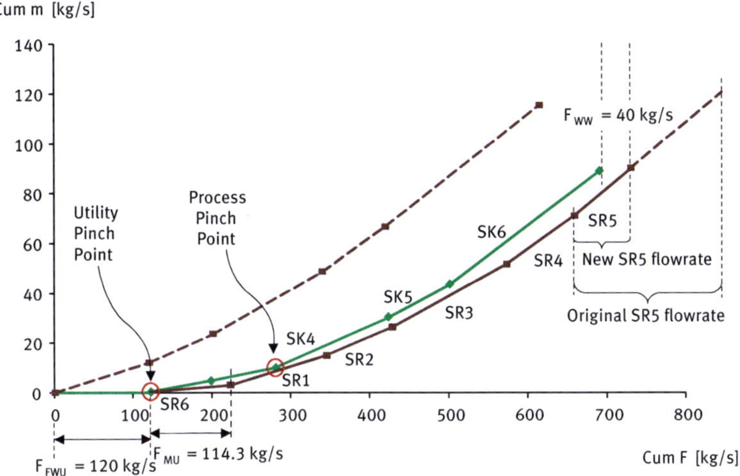

Fig. 6.17: SSCC after SR6 utility addition. SR5 flowrate reduction was exactly the same as the SR6 utility flowrate increment.

Tab. 6.4: Limiting data for Example 6.3 (Wan Alwi and Manan, 2007).

Sink	F, kg/s	C, ppm	m, kg/s	Cum F, kg/s	Cum m, kg/s
SK1	10	0	0	10	0
SK2	120	5	0.6	130	0.6
SK3	50	30	1.5	180	2.1
SK4	80	40	3.2	260	5.3
SK5	50	50	2.5	310	7.8
SK6	30	100	3	340	10.8
SK7	90	150	13.5	430	24.3

Source	F, kg/s	C, ppm	m, kg/s	Cum F, kg/s	Cum m, kg/s
SR1	10	10	0.1	10	0.1
SR2	70	50	3.5	80	3.6
SR3	80	100	8	160	11.6
SR4	50	200	10	210	21.6
SR5	30	300	9	240	30.6
SR6	90	500	45	330	75.6

4. **Example D: Utility addition does not create a Utility Pinch.**

Table 6.4 shows the limiting data for Example 6.3 (Wan Alwi and Manan, 2007). SR7 is a utility added at 130 ppm. Note that, unlike in the previous three examples (Example A to C), shifting SR7 and SLB in this case creates a Process Pinch at $C_{Pinch} = 100$ ppm instead of a Utility Pinch. Fig. 6.18 shows the final Composite Curves after shifting SR7 and SLB along the zero 'cum m' axis line and after moving SLA upwards from the process Pinch along SR7. The F_{MU} for this case is 52.9 kg/s. The Process Pinch Points are at 100 ppm (on line S3) and 200 ppm (on line SR4). The freshwater and wastewater targets are 188.0 kg/s and 140.9 kg/s.

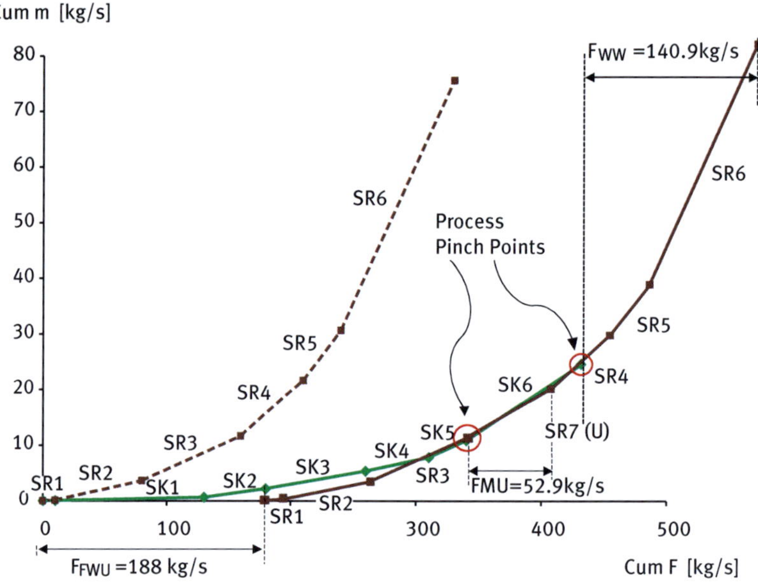

Fig. 6.18: Composite Curves with utility (S7) addition.

6.3.2 Threshold problems

A process which has either freshwater or wastewater as "utility" is a threshold problem. This example focuses on a threshold problem that does not produce wastewater. Note that, since the flowrate of regenerated source equals the flowrate of reduced wastewater source, regeneration would not change the flowrate of freshwater and wastewater for this case. Harvesting external water source provides more room for savings and is therefore the more preferred option. For this type of threshold problem, a Pinch may or may not exist. If a Pinch exists, the utility targeting technique for Pinch problem applies. For threshold problems without a Pinch Point, the following steps should be taken.

1. Shift utility line with SLB until a Utility/Process Pinch Point is created.
2. Shift SLA (SR3) above the Utility/Process Pinch Point upwards along the Utility Line and SLB until all sink flowrates are satisfied.

Table 6.5 is the limiting data for Example 6.4 from Wan Alwi and Manan (2007) and Fig. 6.19 is the corresponding initial SSCC for a threshold problem. The initial freshwater target is 269.99 kg/s. Utility Line (S4) at 80 ppm was added and shifted with SLB (SR1 and SR2) to the right of the Sink Composite until a Pinch Point occurred at $C_{Pinch} = 80$ ppm. SLA (SR3) above the Utility Pinch Point was shifted upwards along the SR4 until the water sink flowrates were fully satisfied (see Fig. 6.20). The F_{MU} for this case is 50 kg/s and the Pinch Point is at 80 ppm. The new freshwater target is 220 kg/s.

Tab. 6.5: Example 6.4 – Limiting data for threshold problem (Wan Alwi and Manan, 2007).

Sink	F, kg/s	C, ppm	m, kg/s	Cum F, kg/s	Cum m, kg/s
SK1	10	0	0	10	0
SK2	120	5	0.6	130	0.6
SK3	130	10	1.3	260	1.9
SK4	50	50	2.5	310	4.4
SK5	30	100	3	340	7.4
SK6	90	200	18	430	25.4
Source					
SR1	10	10	0.1	10	0.1
SR2	70	50	3.5	80	3.6
SR3	80	100	8	160	11.6

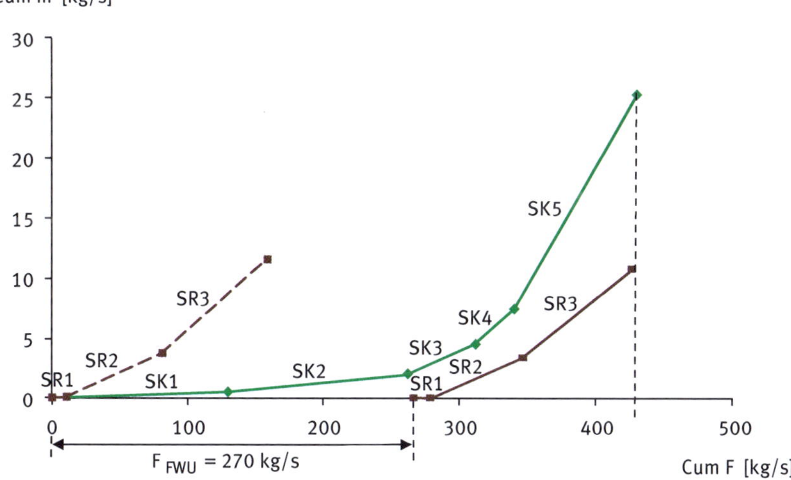

Fig. 6.19: SSCC for Example 6.4 – A threshold problem.

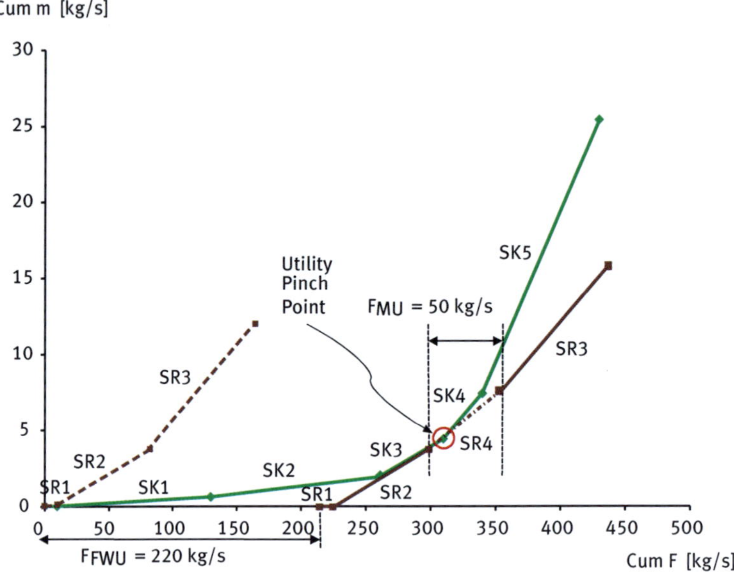

Fig. 6.20: SSCC for threshold problem with addition of SR4 utility.

6.4 Maximum water recovery targets for multiple freshwater sources

A higher quality (cleaner) utility is usually more valuable particularly for cases involving regeneration. Thus, when multiple sources of water and regenerated wastewater are available as utilities, the general rule is to minimise the use of higher quality utility in order to maximise savings. This could be achieved using the following heuristic:

Heuristic 6.2:

Using Water Composite Curves, obtain the F_{MU} one by one, starting from the cleanest to the dirtiest water source.

Heuristic 6.2 means that the F_{MU} for the cleanest *new utility* has to be obtained first using the SSCC procedure described earlier. Adding a utility will create new Utility and Process Pinch points. The next utility could only be considered if its concentration is lower than the highest Pinch concentration. Note that the maximum utility freshwater savings had already been reached with addition of the first utility. Therefore, addition of a dirtier utility should only reduce the flowrate of the cleaner utility added previously. The same procedure is repeated until all available utilities have been utilised. Example E explains how the technique is implemented.

<u>Example E: Two new utilities, *U1* at 10 ppm and *U2* at 80 ppm</u>
To illustrate the technique, the limiting data from Example 6.1 are assessed for possible integration with two new utilities, *U1* and *U2* available at 10 ppm and 80 ppm. Note that *U1* is SR5 from Example A. *U1* as the cleaner utility at 10 ppm is considered first according to Heuristic 1. Fig. 6.21 shows the F_{MU1} for SR5 is 75 t/h and the utility and Process Pinches are at 0 and 100 ppm. *U2* (SR6) at 80 ppm is a viable utility since it existed below the new Pinch Point at 100 ppm. SR6 is drawn next and shifted

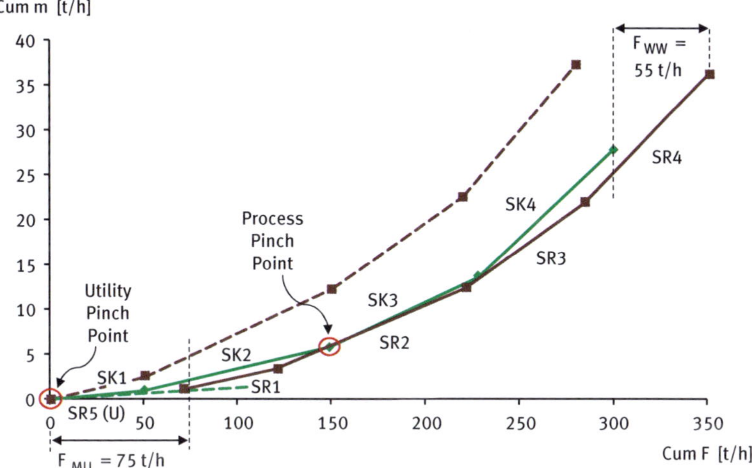

Fig. 6.21: SSCC with addition of U1 (SR5) at C = 10 ppm.

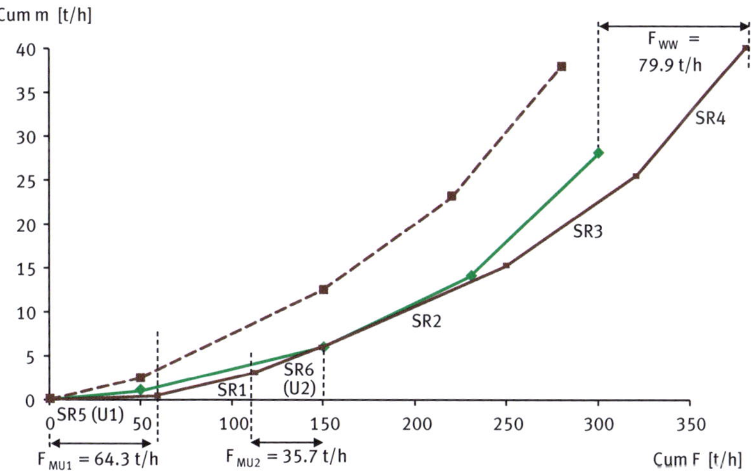

Fig. 6.22: Shifting of SLB and *U2* line (C = 80 ppm) along *U1* (C = 10 ppm) line until a Pinch Point occurred.

with SLB (SR1) downwards along the first utility line (SR5) until another Utility Pinch occurred at 80 ppm (see Fig. 6.22). Next, the SLA above SR6, i.e. SR2 to SR4, are drawn at the new Utility Pinch of 80 ppm and shifted until it completely appeared on the right-hand side of the Sink Composite Curve, or until it created a new Pinch Point. This give F_{MU1} and F_{MU2} of 64.3 t/h and 35.7 t/h (Fig. 6.23). The multiple utility targeting procedures ultimately yield freshwater and wastewater targets at 0 t/h and 79.9 t/h. Note that if SLA had created a new Process Pinch when it was shifted along SR6, any new water source at concentration lower than the new process Pinch Point could still be added to further reduce SR6.

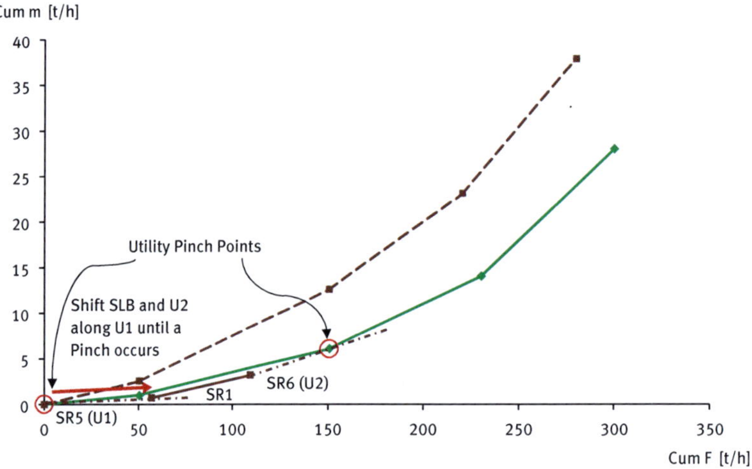

Fig. 6.23: Final SSCC after addition of *U1* and *U2*.

6.5 Working session

Assuming pure freshwater with zero contaminant is available as utility, determine the maximum water recovery target for Examples 5.1 and 5.2 by using:
(a) Water Cascade Analysis (WCA) method
(b) Source/Sink Composite Curve (SSCC) method

6.6 Solution

(a) Water Cascade Analysis (WCA) method
The procedure described in Section 6.2.2 is used to construct the WCT. Table 6.6 shows the WCT for Example 5.1. It can be observed that the largest negative cumulative $F_{FW,k}$ is at 700 ppm, indicating the Pinch location. This value also gives the minimum fresh-

water target of 90.64 t/h. Cascading the flowrate in the last column for the feasible cascade gives the minimum wastewater target of 50.64 t/h.

Tab. 6.6: Water cascade table for Example 5.1.

C_k, ppm	ΔC_k, ppm	ΣF_{SKi}, t/h	ΣF_{SRj}, t/h	$\Sigma F_{SKi} + \Sigma F_{SRj}$, t/h	F_C, t/h	Δm, kg/h	Cum. Δm, kg/h	$F_{FW, cum}$, t/h	F_C, t/h
					0				$F_{FW} = 90.64$
0	−20			−20			0		
	10				−20	−200			70.64
10	−15	10		−5			−200	−20.00	
	90				−25	−2,250			65.64
100	−80	45		−35			−2,450	−24.50	
	100				−60	−6,000			30.64
200	−50			−50			−8,450	−42.25	
	500				−110	−55,000			−19.36
700			50	50			−63,450	−90.64	(PINCH)
	300				−60	−18,000			30.64
1,000			20	20			−81,450	−81.45	
	−1,000				−40	40,000			$F_{WW} = 50.64$

Similarly, Table 6.7 shows the WCT for Example 5.2. The largest negative cumulative $F_{FW,k}$ which is also the Pinch location is at a concentration of 14 ppm. The minimum freshwater target is 2.06 t/h and the wastewater target is 8.16 t/h.

(b) Source/Sink Composite Curve (SSCC) method

For Example 5.1, starting from zero, the sources are first plotted cumulatively in ascending concentration order in a plot of cumulative mass load vs. cumulative flowrate diagram. This gives the Source Composite Curve as shown in Fig. 6.24. The sinks are also plotted cumulatively in ascending concentration order starting from zero. This gives the Sink Composite Curves as shown in Fig. 6.25. The Source Composite Curves are then shifted to the right-hand side of the Sink Composite Curves until a Pinch Point is established as shown in Fig. 6.26. It can be observed that the Pinch Point is located at SR4, which has a concentration of 700 ppm. The minimum freshwater target is the flowrate distance difference between the beginning of the Source Composite Curve and the Sink Composite Curve, i.e. 90.64 t/h. The minimum wastewater target is the flowrate distance difference between the end of the Source Composite Curve and the Sink Composite Curve, i.e. 50.64 t/h.

Tab. 6.7: Water cascade table for Example 5.2.

C_k, ppm	ΔC_k, ppm	ΣF_{SKi}, t/h	ΣF_{SRj}, t/h	$\Sigma F_{SKi} + \Sigma F_{SRj}$, t/h	F_C, t/h	Δm, kg/h	Cum. Δm, kg/h	$F_{FW,cum}$, t/h	F_C, t/h
					0.00				$F_{FW} = 2.06$
0		−1.20	0.80	−0.40			0.00		
	10				−0.40	−4.00			1.66
10		−5.80		−5.80			−4.00	−0.40	
	4				−6.20	−24.80			−4.14
14			5.00	5.00			−28.80	−2.06	(PINCH)
	11				−1.20	−13.20			0.86
25			5.90	5.90			−42.00	−1.68	
	9				4.70	42.30			6.76
34			1.40	1.40			0.30	0.01	
					6.10	0.00			$F_{WW} = 8.16$

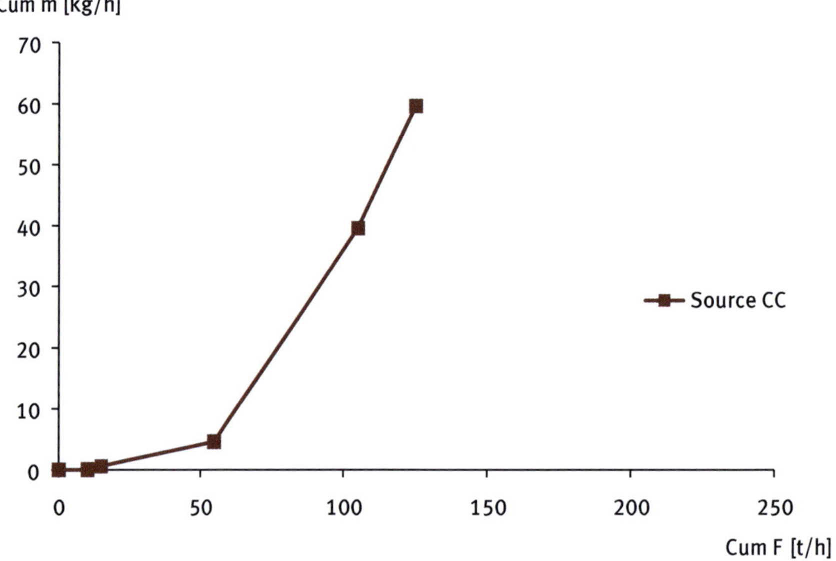

Fig. 6.24: Source Composite Curve for Example 5.1.

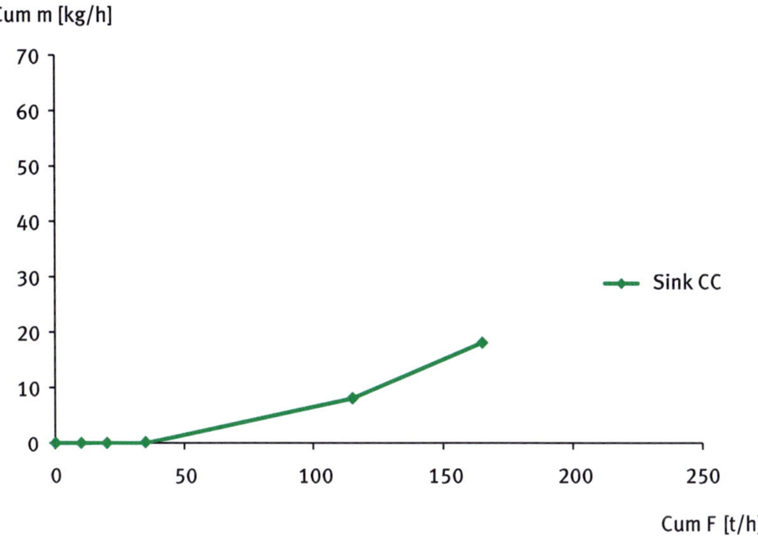

Fig. 6.25: Sink Composite Curve for Example 5.1.

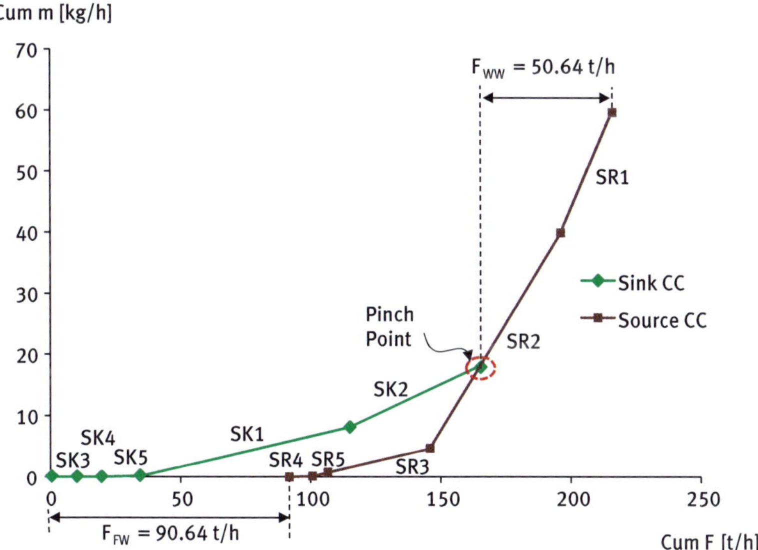

Fig. 6.26: Source/Sink Composite Curve for Example 5.1.

Similar procedures are performed for Example 5.2. The final SSCC is shown in Fig. 6.27. It can be observed that the Pinch Point is also located at SR2, 14 ppm. The minimum freshwater target is 2.06 t/h and the minimum wastewater target is 8.16 t/h.

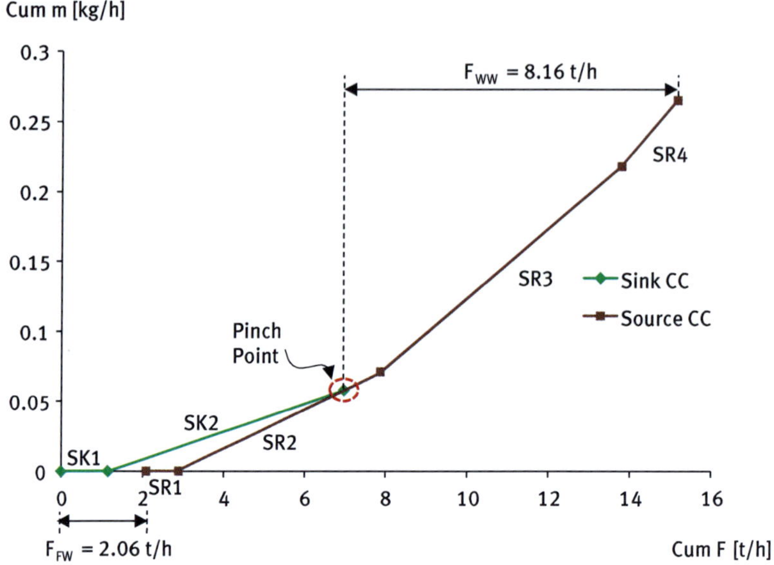

Fig. 6.27: Source/Sink Composite Curve for Example 5.2.

References

Almutlaq, A.M., Kazantzi, V. and El-Halwagi, M. (2005). An algebraic approach to targeting waste discharge and impure fresh usage via material recycle/reuse networks, *Cleaner Technology and Environmental Policy*, 7(4), 294–305.

Bandyopadhyay, S. (2006). Source Composite Curve for waste reduction, *Chemical Engineering Journal*, 125(2), 99–110.

El-Halwagi, M.M., Gabriel, F. and Harell, D. (2003). Rigorous graphical targeting for resource conservation via material recycle/ reuse networks, *Ind. Eng. Chem. Res.*, 42, 4319–4328.

Foo, D.C., Manan, Z.A. and Tan, Y.L. (2005). Synthesis of maximum water recovery network for batch process systems, *J. Cleaner Production*, 13(15), 1381–1394.

Foo, D.C., Manan, Z.A. and Tan, Y.L. (2006). Use cascade analysis to optimize water networks, *Chem. Eng. Progress*, 102(7), 45–52.

Hallale, N. (2002). A new graphical targeting method for water minimization, *Advances in Environmental Research*, 6(3), 377–390. DOI: 10.1016/S1093–0191(01)00116–00112.

Majozi, T., Brouckaert, C.J. and Buckley, C.A. (2006). A graphical technique for wastewater minimisation in batch processes, *Journal of Environmental Management*, 78, 317–329.

Manan, Z.A., Tan, Y.L. and Foo, D.C.Y. (2004). Targeting the minimum water flowrate using water cascade analysis technique, *AIChE Journal*, 50(12), 3169–3183.

Polley, G.T. and Polley, H.L. (2000). Design better water networks, *Chemical Engineering Progress*, 96(2), 47–52.

Prakash, R. and Shenoy, U.V. (2005). Targeting and design of water networks for fixed flowrate and fixed contaminant load operations, *Chemical Engineering Science*, 60(1), 255–268. DOI: 10.1016/j.ces.2004.08.005.

Sorin, M. and Bedard, S. (1999). The global Pinch point in water reuse networks, *Process Safety Environ. Prot.*, 77, 305–308.

Wan Alwi, S.R. and Manan, Z.A. (2007). Targeting multiple water sources using Pinch Analysis, *Industrial & Engineering Chemical Research*, 46(18), 5968–5976.

Wang, Y.P. and Smith, R. (1994). Wastewater minimisation, *Chemical Engineering Science*, 49(7), 981–1006.

Wang, Y.P. and Smith, R. (1995). Wastewater minimization with flowrate constraints, *Chem. Eng. Res. Des.*, 24, 2093–2113.

7 Water network design/retrofit

7.1 Introduction

Once the maximum water recovery target has been determined, the next step is to design a water network which can achieve this target. Two established method, i.e. the Source/Sink Mapping Diagram (SSMD) and the Source and Sink Allocation Curves (SSAC) will be described in this chapter.

7.2 Source/Sink Mapping Diagram (SSMD)

The SSMD was introduced by Polley and Polley (2000). The authors proposed a design principle that involves the systematic mixing of source streams with each other and, where necessary, with freshwater to produce streams that just meet the individual sinks in terms of both quantity and quality. The quantity of water to be sent to the contaminated source is determined by calculating the contaminant load associated with the sink. The result takes one of the forms:

(a) If this quantity is equal to the required value, then no freshwater needs to be supplied.
(b) If the quantity is less than the sink, then some freshwater is required.
(c) If the quantity required from the source exceeds the sink, then part of the sink's contaminant load should be satisfied using a source having a higher contaminant concentration.

Constraints for network design between water sources i and sink j are given as follow (Hallale, 2002):

(a) Sinks
   1. Flowrate

$$\sum_i F_{i,j} = F_{SK,j}.$$

(7.1)

    Where F_i is the total flowrate available from source i to Sink j. Where $F_{SK,j}$ is the flowrate required by Sink j.

   2. Concentration

$$\frac{\sum_i F_{SRi,SKj}\, C_i}{\sum_i F_{SRi,SKj}} \leq C_{\max,SKj}.$$

(7.2)

    Where C_{SKj} is the contaminant concentration of source i and $C_{max,SKj}$ is the maximum acceptable contaminant concentration of sink j.

(b) Sources

1. Flowrate

$$\sum_i F_{SRi,SKj} \leq F_{SRi}. \tag{7.3}$$

This network design method can stand on its own even without the targeting stage. However, for some cases, the network design is not the same as in the targeting stage since the cleanest to cleanest rule does not hold for the mass load deficit case (this will be explained further in Section 7.3). If the designers would like to use this method, it is important to ensure that the resulting freshwater and wastewater flowrates match the minimum water targets.

Example 6.1 from the previous chapter will be used to illustrate this method. All the water sinks are aligned horizontally at the top while all water sources are lined up vertically on the left-hand side. Both are arranged according to increasing contaminant concentration. All the water flowrates and mass loads for each source and sink are listed. Pinch regions are also labelled on the diagram (see Fig. 7.1).

Start the source-sink matching process by matching the sink with the lowest contaminant concentration with the source with the lowest contaminant concentration. The quantity of water from the contaminated source required is determined by the quality (contaminant load) and quantity (flowrate) of the sink. For example, SK1 requires 50 t/h and can accept a mass load up to 1,000 kg/h. SR1 has a flowrate of 50 t/h and a mass load of 2,500 kg/h. Hence, only a flowrate of 20 t/h, which corresponds to 1,000 kg/h of mass load, can be used from SR1 to satisfy SK1 quality requirement. The remaining 30 t/h of flowrate required by SK1 needs to be satisfied by using freshwater which has zero mass load (assuming freshwater does not have any contaminants). This will satisfy both the quantity and quality required by the SK1.

Next, SK2 is matched with the remaining SR1. Since SK2 requires a water flowrate of 100 t/h and a mass load of 5,000 kg/h, all the remaining SR1 (30 t/h of flowrate with 1500 kg/h of mass load) can be sent to SK2. Since SK2 quantity and quality has still not been satisfied yet (still requires F = 70 t/h and m = 3,500 kg/h), the next source, i.e. SK2 is considered. 35 t/h of SK2 with 3,500 kg/h mass load are added to SK2 to satisfy the quality needed. The remaining flowrate of SK2 is satisfied by using freshwater (35 t/h). This step is repeated for the remaining sinks.

Note that below the Pinch region, both the quantity and quality of the sink must be satisfied by either the sources or in combination with freshwater. However, for above the Pinch region, only the quality of the sink needs to be satisfied while the quantity can be equal or less. The excess water source, when all water sink have been matched, are sent to the wastewater treatment plant. Since the freshwater and wastewater are the same as those targeted in the previous chapter, hence this design is correct. Fig. 7.1 shows the completed SSMD.

Fig. 7.1 shows one possible network design generated by the SSMD. Note that this is only one of many possible design that can achieve the targets. The designers can influence the solution by imposing other constraints such as forbidden or forced connection for safety or economic reasons. However, these additional constraints may sometimes result in a water penalty.

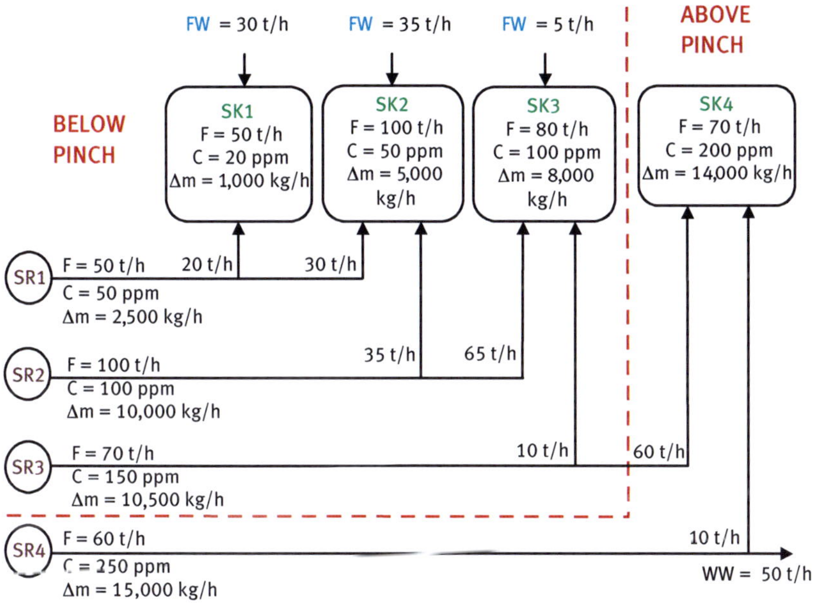

Fig. 7.1: Network design by Source Sink Mapping Diagram (Polley and Polley, 2000).

7.3 Source and Sink Allocation Curves (SSAC)

The SSAC can be used to show the water network allocation or design between the source and the sink. SSAC was initially proposed by El-Halwagi et al. (2003) for pure freshwater sources and Kazantzi and El-Halwagi (2005) for impure freshwater sources. Both methods are based on the cleanest to cleanest rule, similar to the SSMD method by Polley and Polley (2003). All the three methods are however limited to the *'flowrate deficit case'*, i.e. the case where the mass load of a sink is satisfied but not its flowrate, as shown in Fig. 7.2. There is another important case which is the *'mass load deficit case'*, where the source(s) meets only the flowrate of a sink but not the mass load (see Fig. 7.2(b)). The cumulative water sources below and between Pinch regions must have the exact same flowrate and mass load as the cumulative sink. Note that using the cleanest source that does not meet the mass load of a sink results in excess mass load of source and ultimately unsatisfied flowrate of subsequent sinks.

Hence, Wan Alwi and Manan (2008) introduced another rule for the 'mass load deficit case', which is to satisfy the mass load and flowrate of a sink using the cleanest as well as the dirtiest sources in the Pinch region and making sure the Pinch point and the utility targets are satisfied. The authors also introduced the Network Allocation Diagram (NAD) which translates the Source and Sink Allocation Composite Curves into a network mapping diagram for easy visualisation.

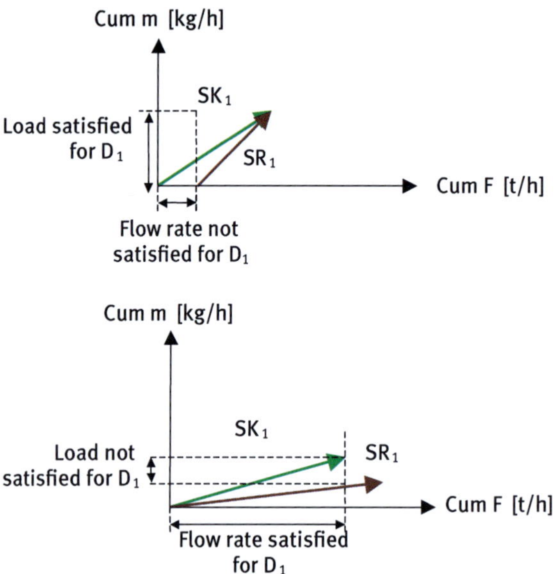

Fig. 7.2: Satisfying the cleanest sink with cleanest water source using SSAC (a) flowrate deficit case, (b) mass load deficit case (Wan Alwi and Manan, 2008).

The method by Wan Alwi and Manan (2008) comprises three steps:

Step 1: Plot the sources and sinks from the cleanest to the dirtiest cumulatively to form a Source/Sink Composite Curve (SSCC) as described in Chapter 6. Note the minimum freshwater and wastewater flowrate. Note also the region where the sources and sinks are located.

Step 2: Draw the Source and Sink Allocation Curves (SSAC) for the two separate Pinch regions based on the following rules:
(a) Matching sources and sinks in the region below the Pinch and between Pinches
1.   Begin by matching the cleanest source with the cleanest sink, and pick the subsequent matches as follows:

2. For the case where freshwater or utility purity is superior to *all* other streams (see Fig. 7.3):

    (i)   If the cumulative water source(s) satisfies the mass load but not the flowrate as shown in Fig. 7.2(a), add water utility until the sink flowrate is satisfied (only valid for below Pinch region). Once both the flowrate and the mass load of the sink have been satisfied, match the remaining sources with the next sink.

    (ii) If the cumulative water source(s) satisfies the flowrate but not the mass load as shown in Fig. 7.2(b), pick the dirtiest source in this Pinch region to satisfy the remaining mass load of the first sink. Once both the flowrate and the mass load of the first sink have been satisfied, use the remaining dirtiest source to satisfy the mass load of the next sink.

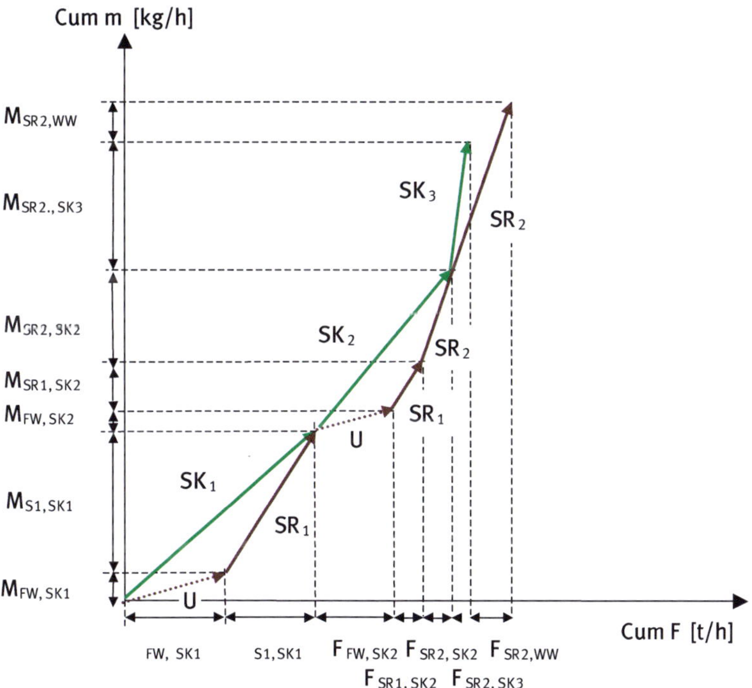

Fig.7.3: Source and Sink Allocation Curves with utility stream concentration superior than all other streams.

3. For the case where freshwater or utility purity is not superior to *all* other streams (see Fig. 7.4) (this rule can also be applied for the availability of utility superior to other streams):

    (i)   Treat the utility as a one of the water sources.

(ii) Satisfy the mass load and flowrate of a sink by using the cleanest and the dirtiest sources in this Pinch region. Note the minimum amount of utility targets that have already been established in Step 1.

(iii) Once both the flowrate and mass load of the sink have been satisfied, match the remaining sources with the next sink.

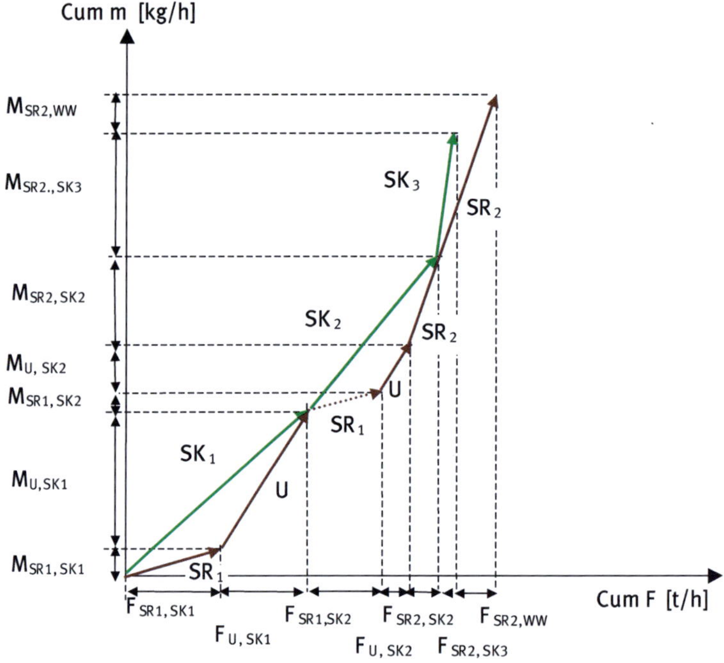

Fig. 7.4: Source and Sink Sllocation Curves with utility stream concentration not superior than all other streams.

(b) Matching sources and sinks in the region above the Pinch. Note that the region above the Pinch refers to the starting point of the last Pinch Point to the point before the water source(s) becomes wastewater.

4. The sources above the Pinch are used in ascending concentration order to satisfy the sink flowrate requirements. Note that no freshwater or higher concentration source shifting is required above the Pinch. All sources' mass loads above the Pinch are less than mass loads of sinks. The flowrate provided by the sources is sufficient for the sinks but the mass load will be less. This means that sinks above the Pinch will be fed by a higher purity source than required.

5. If there is no more sinks to satisfy, the remaining sources will become wastewater.

Step 3: Draw the Network Allocation Diagram (NAD) based on the SSAC. The exact flowrate of each source allocated to each sink can be obtained directly from the length of the x-axis.

7.3.1 Example of network design using SSCC for utility purity superior to all other streams

Example 6.2 will be used to illustrate the network design step for the case where utility purity is superior to all other streams. In this case, utility has a concentration of 0 ppm.

Step 1: Plot the sources and sinks from the cleanest to the dirtiest cumulatively to form a Source and Sink Composite Curve (SSCC).

The water sink and source lines are first plotted according to increased concentration. A locus of utility line is then drawn starting from the origin, with the slope represented by the utility concentration. The water source line is then shifted to the right along the utility line (Kazantzi and El-Halwagi, 2005) until all the water source lines are on the right-hand side of the sink line. The point where the water source line touches the water sink line is the Pinch Point. The new utility flowrate is given by the length of horizontal utility line after shifting the water source lines to the right. The minimum wastewater target is the overshoot of the source line. The SSCC for Example 2.2 are as shown in Fig. 7.5. The Pinch Points are noted at 100 ppm and 180 ppm. This defines the region below, between and above Pinch. The freshwater and wastewater targets are 200 t/h and 120 t/h.

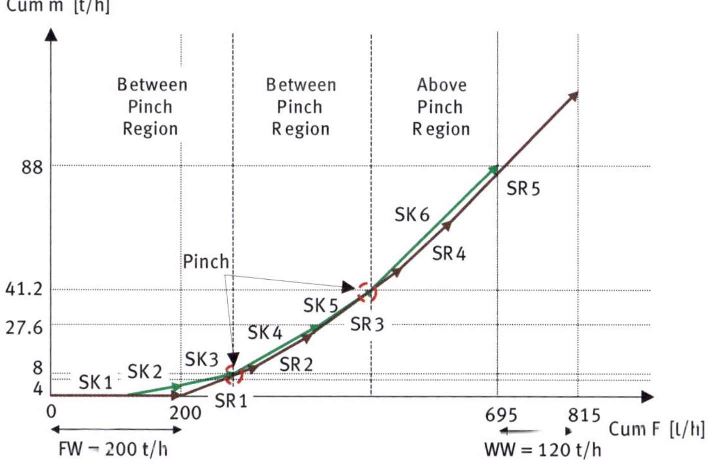

Fig. 7.5: Source/Sink Composite Curve for Example 6.2.

Step 2: Draw the SSAC based on the proposed rules.

1.   Region below the Pinch

There are three sinks located below the Pinch region, i.e. SK1, SK2 and SK3 (see Fig. 7.5). SK1 required 120 t/h of water with zero mass load. Since there was no water source with zero mass load, hence the first sink flowrate was completely satisfied using freshwater. This corresponded to 120 t/h of freshwater usage.

For SK2, only part of SR1 could satisfy the mass load of SK2 as shown in Fig. 7.6. As stated in *Rule 2(i)*, if the cumulative water source(s) satisfies the mass load but not the flowrate, add water utility until the sink flowrate is satisfied. Freshwater was added until the entire sink flowrate was satisfied. Thus, 40 t/h of SR1 and 40 t/h of freshwater are needed to satisfy SK2.

The final sink located below the Pinch region was SK3. The remaining SR1 located below the Pinch region was used to satisfy SK3 mass load requirement. Again, *Rule 2(i)* applied for this case. Freshwater was added until the entire sink flowrate requirements were satisfied. Hence, 40 t/h of SR1 and 40 t/h of freshwater were fed to SK3. The whole process of the source and sink allocation below the Pinch region is shown in Fig. 7.6.

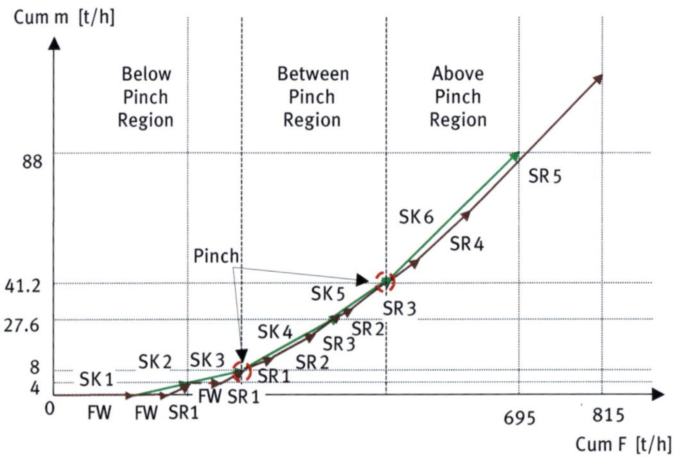

Fig. 7.6: Final SSAC for Example 6.2.

2.   Region between Pinches

There were two sinks located between the Pinch region i.e. SK4 and SK5 as illustrated in Fig. 7.5. SK4 flowrate was satisfied using part of SR1, all of SR2 and part of SR3 in ascending order of source concentration based on *Rule 1*. However, as shown in Fig. 7.7, the flowrate of SK4 has already been satisfied but not the mass load requirement. Using *Rule 2(ii)*, SR3 as the dirtiest line in the region was shifted downwards along the SR2 line until all the mass load and flowrate requirements of SK4 were satisfied. Hence 40 t/h of S1, 60 t/h of SR2 and 40 t/h of SR3 were used for SK4. The remaining SR2 and SR3 corresponding to 20 t/h and 60 t/h respectively in the Pinch region were used to satisfy the remaining sink, i.e. SK5. Fig. 7.6 shows the final source and sink allocation for the region between Pinches.

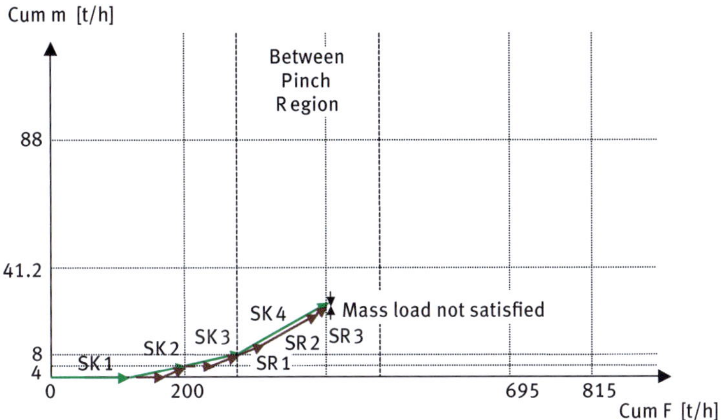

Fig. 7.7: SSAC for Example 6.2 where SK4 mass load was not satisfied for the region between Pinches.

3. Region above the Pinch

Finally, to satisfy the flowrate for the above Pinch region, the Pinch sources in the region were used in ascending order to satisfy the sink flowrate requirement (*Rule 4*). Hence, 40 t/h of SR3, 80 t/h of SR4 and 75 t/h of SR5 were used to satisfy SK4 flowrate. The amount of mass load cumulated from these three sources only totalled 44.35 kg/h, but SK4 could actually accept 48.75 kg/h. The remaining SK5 was rejected as wastewater since there was no more sink left to satisfy (*Rule 5*). The above Pinch source and sink allocation are as shown in Fig. 7.6.

Step 3: Draw the Network Allocation Diagram (NAD) using the SSACs.

Now that Steps 1 and 2 are completed, the Network Allocation Diagram (NAD) can be drawn. The x-axis was segmented into sink flowrate intervals by drawing vertical lines. All the water sinks were aligned horizontally at the top while all water sources are lined up vertically on the left-hand side. Both axes were arranged according to increasing contaminant concentration. Note that the lengths of water sinks were drawn to match the sink flowrate obtained from the final source and sink allocation curves.

The x-axis was further segmented into source flowrate intervals by drawing vertical lines. The sources were then drawn horizontally just below the water sinks according to the SSAC. This represented the exact amount of source allocated for the respective sink. Finally, arrows were drawn linking the vertically aligned sources to the horizontally aligned sources to show how the sources were distributed into the sinks.

The final network diagram is as in Fig. 7.8. Note that without the SSAC procedure as a guide, a network diagram may miss the minimum freshwater and wastewater targets for certain cases that were designed using the cleanest to cleanest rule proposed by Polley and Polley (2000).

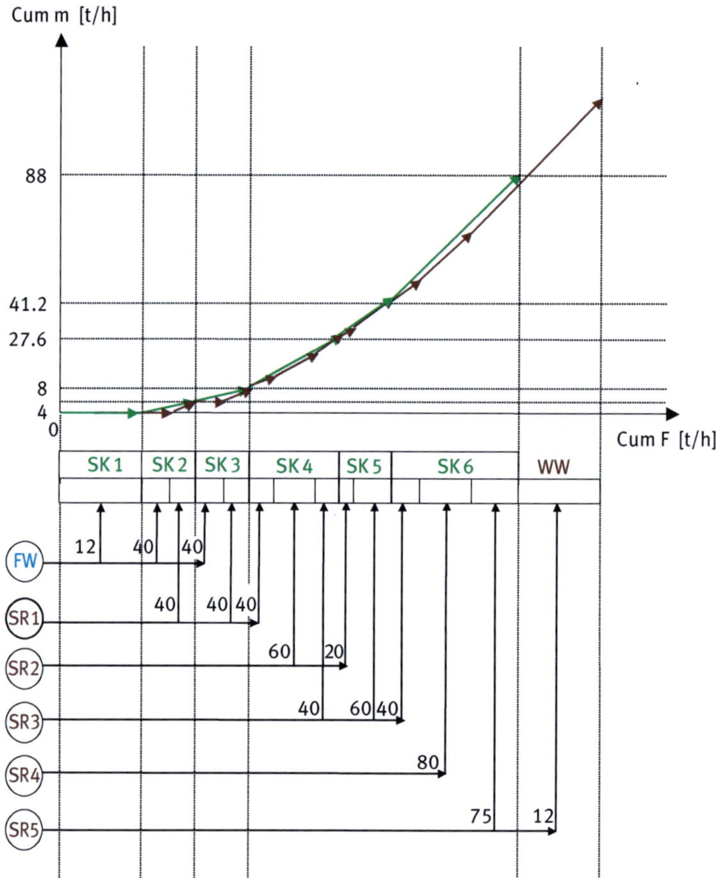

Fig. 7.8: Final Network Allocation Diagram.

7.3.2 Freshwater purity not superior to all other streams

Table 7.1 shows the limiting data for Example 7.1 to illustrate the case where freshwater purity is not superior to all other streams. In this case, freshwater has a concentration of 30 ppm. A real-life example for cases that some sinks or sources exist at a concentration below freshwater concentration is a semiconductor plant where ultra-pure water is used widely and the wastewater produced is cleaner than freshwater.

Tab. 7.1: Limiting data for Example 7.1.

Sink	F, t/h	C, ppm	m, t/h
SK1	70	20	1.4
SK2	70	30	2.1
SK3	100	120	12
Source			
SR1	50	10	0.5
SR2	75	50	3.75
SR3	100	150	15

Step 1: Plot the sources and sinks from the cleanest to the dirtiest cumulatively to form an SSCC.

Drawing the SSCC, cumulative water sink lines are plotted first in ascending order of concentration. The cumulative water source(s) that have concentrations lower than the freshwater concentration are plotted next, corresponding to SR1. A locus of utility line is then drawn starting from the end of SR1 line, with a slope of 30 ppm. The remaining water source line (also arranged in ascending order), that is SR2 and SR3, are then shifted to the right along the utility line until all the water source lines are on the right-hand side of the sink line. The point where the water source line touches the water sink line is the Pinch Point. The new utility flowrate is given by the length of horizontal freshwater line after shifting the remaining water source lines to the right. The minimum wastewater target is the overshoot of the source line. The Source and Sink Composite Curves for Example 7.1 are as shown in Fig. 7.9. The Pinch Point is noted at 50 ppm. This defines the region below and above the Pinch. The freshwater and wastewater targets are 75 t/h and 60 t/h respectively.

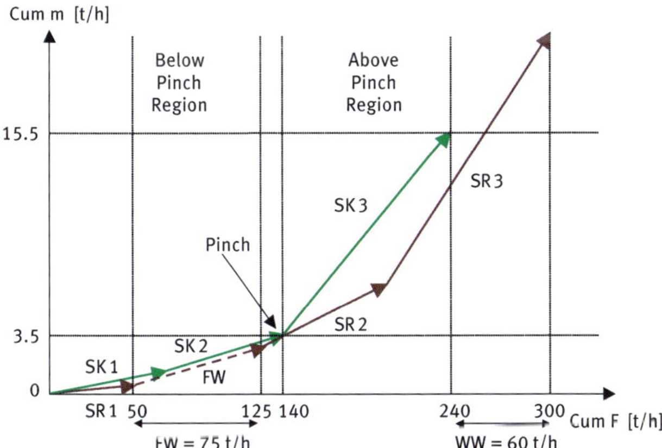

Fig. 7.9: Source and Sink Composite Curves for Example 7.1.

Step 2: Draw the SSAC based on the proposed rules.

1. Region below the Pinch

There were two sinks located below the Pinch region, i.e. SK1 and SK2 (from Fig. 7.9). SK1 required 70 t/h of water with 1.4 kg/h mass loads. Using *Rule 3(i)*, utility was treated as a source as well since the utility concentration was not superior to all other streams. Using *Rule 1*, the cleanest cumulative source was used first to satisfy the flowrate and mass load of SK1. It can be seen from Fig. 7.10 that the use of all SR1 and part of the utility or freshwater line satisfy only the flowrate but not the mass load of SK1. Hence, based on *Rule 3(ii)*, the dirtiest source in the region below the Pinch was used. Below the Pinch region, SR2 line was shifted downward along freshwater and SR2. Hence, 50 t/h, 5 t/h and 15 t/h of SR1, freshwater and SR2 were needed to satisfy both SK1 flowrate and mass load. The remaining freshwater line at 70 t/h flowrate was then used to satisfy SK2 as *Rule 3(iii)* stated. The final SSAC for the region below the Pinch for Example 3.1 is as shown in Fig. 7.11.

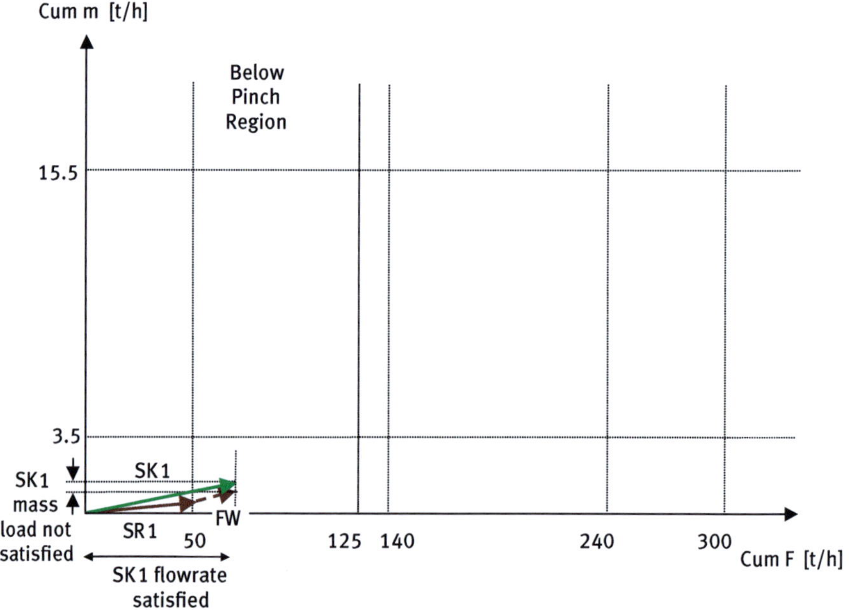

Fig. 7.10: SSAC for Example 7.1 using Rule 1 satisfying SK1 for region below Pinches.

3. Region above the Pinch

Finally, to satisfy the above Pinch regions flowrate, the Pinch sources in the region were used in ascending order to satisfy the sink flowrate requirement (*Rule 4*). Hence, 60 t/h of SR2 and 40 t/h of SR3 were used to satisfy SK3 flowrate. The amount of mass load cumulated from these two sources only totalled 9 t/h but SK3 could actually accept 12 t/h. The remaining 60 t/h of SR3 were rejected as wastewater since there was no more sink to be satisfied (*Rule 5*). The above Pinch source and sink allocation are as shown in Fig. 7.11.

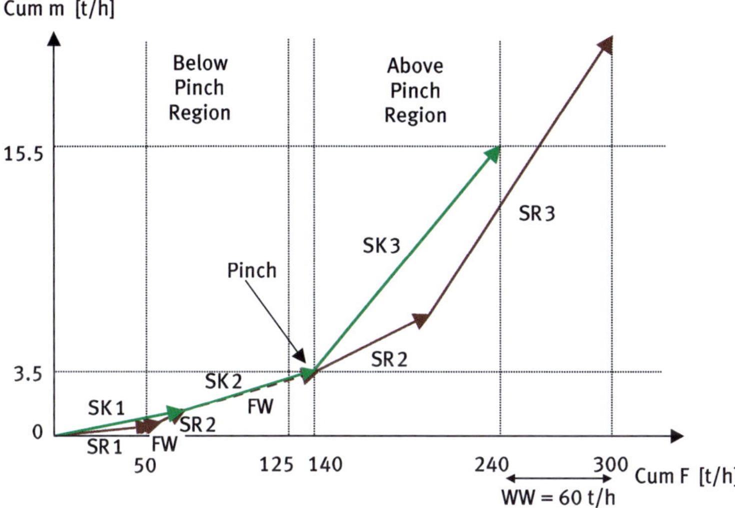

Fig. 7.11: Final SSAC for Example 7.1.

Step 3: Draw the NAD based on the SSAC.

The NAD was drawn upon completion of Steps 1 and 2. The final network diagram for Example 7.1 is as in Fig. 7.12.

7.3.3 Simplification of a water network or constructing other network possibilities

The network diagram obtained using the step-wise procedure described previously was only one of the many possible network designs that could achieve the minimum freshwater and wastewater targets. The network allocation proposed however yielded the minimum freshwater and wastewater targets that resulted in a complex network due to the many mixings of streams and might pose operability problems associated with geographical constraints, safety and cost. The SSAC from Step 2 can actually be further used to explore various other network possibilities in order to reduce the network complexity. The rules for other various source and sink allocation possibilities are as follows:

6. *For allocating source and sink without water penalty, shift, cut and allocate the sources line to satisfy any sink in the same Pinch region provided that the sink mass load and flowrate requirements are fully satisfied.*

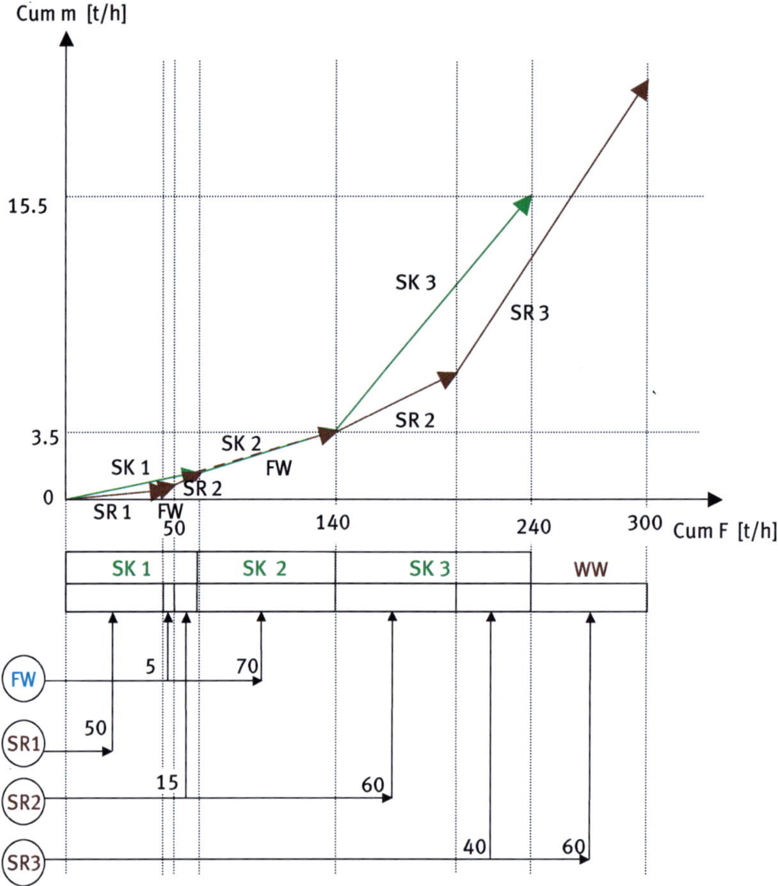

Fig. 7.12: Final Network Allocation Diagram for Example 7.1.

7. *For allocating source and sink with water penalty, satisfy a sink using the minimum number of source streams available either in the same Pinch region or from other Pinch regions. A penalty of increased freshwater consumption will be incurred.*

Rule 6 applies for other source and sink allocations that still achieve the minimum freshwater and wastewater targets. *Rule 7* applies for source and sink allocations that yield a simpler water network but result in a penalty of increased freshwater consumption.

For example, Fig. 7.13 is the SSAC constructed using Step 2 from the previous section for Example 2.1 limiting water data. Using *Rule 6*, Fig. 7.14 is another possible SSAC that achieved the minimum freshwater and wastewater targets in addition to Fig. 7.13. In this case only sources below the Pinch region were cut and shifted around. For instance, part of SR2 ($F_{SR2,SK1} = 10$ t/h) was cut and shifted for use with SK1 instead of SK2 initially. This caused the FW needed to satisfy SK1 to increase to 40 t/h. Next, the network was simplified by shifting all SR1 ($F_{SR1,SK2}$ = 50 t/h) for use only with SK2, with part of SR2 ($F_{SR2,SK2}$ = 10 t/h) and SR3 also cut and shifted ($F_{SR3,SK2}$ = 10 t/h). All the remaining FW were used to satisfy SK2 ($F_{FW,SK2}$ = 30 t/h) and all SK3 was satisfied using the remaining SR2 ($F_{SR2,SK3}$ = 80 t/h).

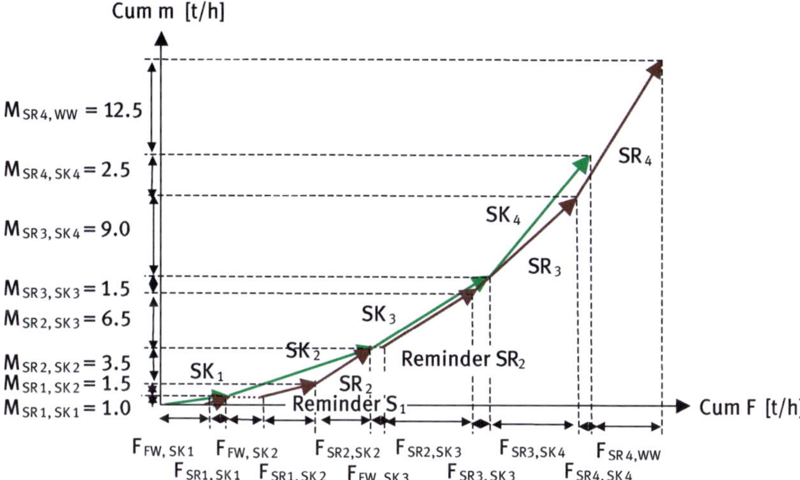

Fig. 7.13: The SSAC using Step 2 for Example 6.1.

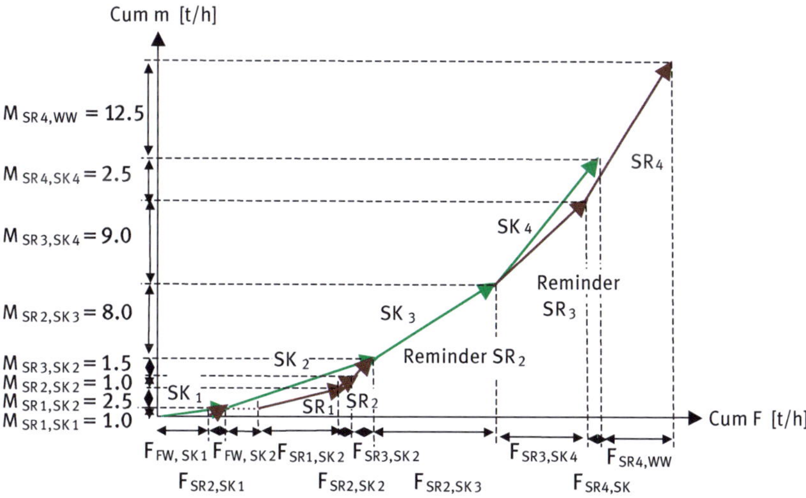

Fig. 7.14: Another possible SSAC achieving the same minimum freshwater and wastewater flowrate targets.

Fig. 7.15 is a possible SSAC based on *Rule 7* that yielded a simpler structure with fewer splits. However, penalties for freshwater and wastewater were incurred, with the new freshwater target of 100 t/h and 80 t/h compared to the initial targets of 70 t/h and 50 t/h respectively. The sources from above and below the Pinch were used in any region to satisfy the sink flowrate but not necessarily the sink mass load. The sink was either fed with sources that had the same total mass load as the sink needed or lesser. From Fig. 7.15 it can be seen that:

- All SK1 are fulfilled using freshwater.
- All SR1 are used to satisfy SK2 and the remaining flowrate of SK2 was satisfied using freshwater.
- All SR2 are used to satisfy SK3 and the remaining SR2 sent to wastewater.
- All SR3 are used for SK4 and the remaining SR4 becomes wastewater.

Based on the two examples, it can be clearly seen that SSAC could provide a powerful guideline for network modifications based on the two proposed rules. A designer could opt to design a network with or without water penalty using the SSAC. Once satisfied with the SSAC, the designer could proceed to network design using Step 3.

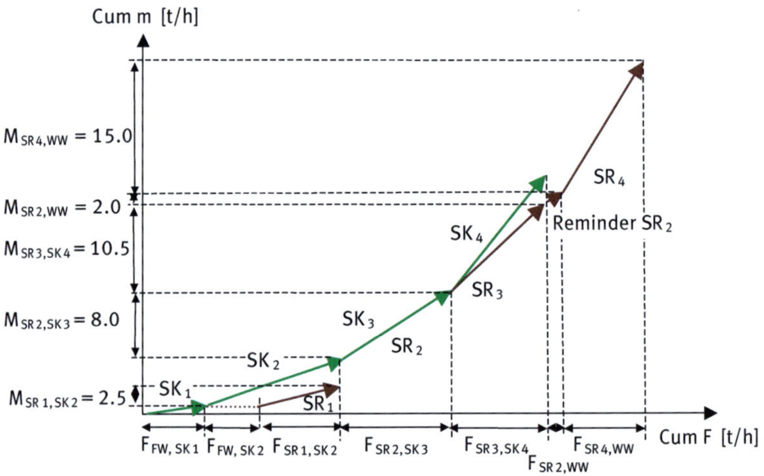

Fig. 7.15: A possible SSAC with freshwater and wastewater penalty.

7.4 Working session

As an assignment, construct the water network design for Example 5.1 by using SSMD and SSAC method.

7.5 Solution

1. Source/ and Sink Mapping Diagram

Fig. 7.16 shows the completed Source/ and Sink Mapping Diagram for Example 5.1. The sources are arranged in ascending order vertically on the left while the sinks are arranged in ascending order horizontally on the top.

SK3 and SK4 are both satisfied by using freshwater since it requires mass load of 0 t/h and there are no available sources with Δm = 0 t/h. SK5, which requires Δm of 0.15 t/h, is satisfied by using 10 t/h of SR4 (Δm = 0.1 t/h) and 0.5t/h of SR5 (Δm = 0.05 t/h). The remaining flowrate of 4.5 t/h required by SK5 is satisfied by using freshwater. SK1, which requires Δm of 8 t/h, is satisfied by using 4.5 t/h of SR5 (Δm = 0.45 t/h), 40 t/h of SR3 (Δm = 4 t/h) and 5.07 t/h of SR2 (Δm = 3.55 t/h). The remaining required SK1 flowrate of 30.43 t/h is satisfied by using freshwater.

SK2, which requires Δm of 10 t/h, is satisfied with 14.29 t/h of SR2 (Δm = 10 t/h). The remaining flowrate of 35.71 t/h required by SK2 is satisfied by using freshwater. Since all sinks have been satisfied, the remaining 30.64 t/h of SR2 and 20 t/h of SR1 are sent to the wastewater treatment plant. The summation for all freshwater is 90.64 t/h

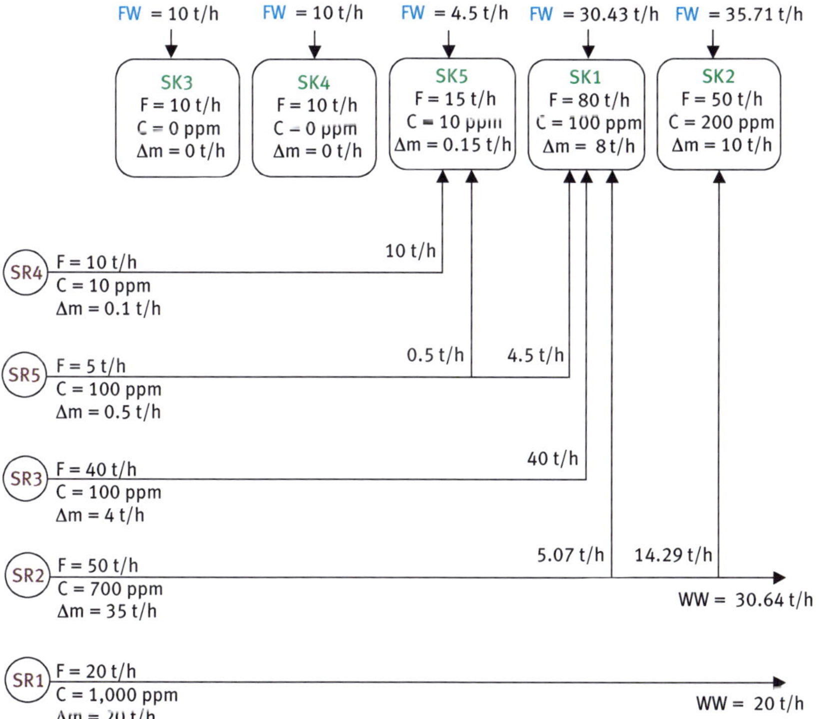

Fig. 7.16: Source and Sink Mapping Diagram for Example 5.1.

and for all wastewater is 50.64 t/h. This is similar to the targeted value in Chapter 6 and hence the network design is correct.

2. Source and Sink Allocation Curve

The minimum freshwater and wastewater targets are first targeted by using the SSCC as described in the previous chapter. The Pinch location is noted to be at SR2 with concentration of 700 ppm. Next, SSAC is constructed for Example 5.1 by using the methodology described in Section 7.3. Below Pinch Region SSAC are performed first. For SK3 and SK4, freshwater is used since there are no sources with zero mass load. SK5 is matched with SR4 and part of SR5. Since this is a *flowrate deficit case*, therefore, freshwater is added until the sink flowrate is satisfied. The remaining sinks are satisfied with the remaining sources. Since all source and sink matches are *flowrate deficit cases*, freshwater is added in order to satisfy the flowrate requirements. For the

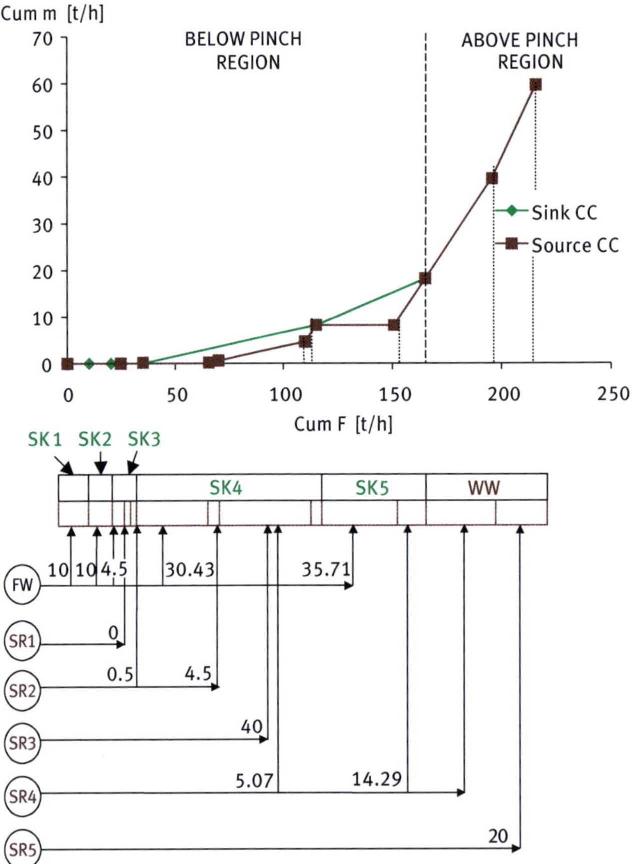

Fig. 7.17: Source/Sink Allocation Diagram and Network Allocation Diagram for Exercise 5.1.

above Pinch region, since there are no more sinks, all the remaining water sources are sent to the wastewater treatment plant.

The SSAC results are then directly translated into NAD, which looks almost similar to SSMD, but with the length of the sources and sinks representing the actual amount of flowrate transferred. Fig. 7.17 shows the SSAC and NAD for Example 5.1. The freshwater and wastewater targets are similar to the previous method.

7.6 Water MATRIX software

Optimal Water© (2006) (formerly known as Water MATRIX©) is software developed by Process Systems Engineering Centre (PROSPECT), Universiti Teknologi Malaysia. The software can target and design the maximum water recovery network using the Water Pinch Analysis technique. It incorporates the Balanced Composite Curves, Water Surplus Diagram, Water Cascade Analysis and Source/Sink Mapping Diagram methods. User only needs to key in the sources' and sinks' flowrates and contaminant concentrations. The software then automatically generates the BCC, WSD, WCA and SSMD. Figs. 7.28 and 7.29 show some of the software's features.

Conc, C (ppm)	ΔP	Sum F Source (kg/s)	Total F (kg/s)	Cum. water flowrate (kg/s)	Cum. water surplus (kg/s)
				373.296	
0			0		
	0.00002			373.296	
20			-466.62		0.00746592
	0.00008			-93.324	
100		201.84	201.84		0
	0.00005			108.516	
150		1,131.54	1,131.54		0.0054258
	0.00002			1,240.056	
170		390.96	-860.88		0.03022692
	0.00005			379.176	
220		265.2	265.2		0.04918572
	0.00006			644.376	
280			-68.7		0.08784828
	0.00002			575.676	
300		68.7	68.7		0.0993618
	0.9997			644.376	
					644.282049

Fig. 7.18: Water MATRIX – Water Cascade Table.

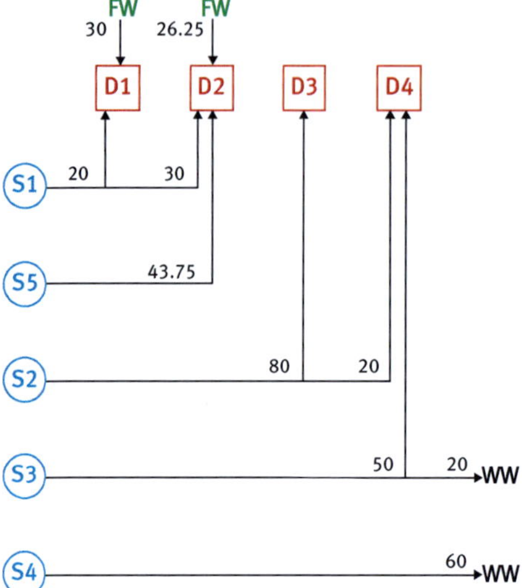

Fig. 7.19: Water MATRIX – Source/Sink Mapping Diagram.

References

El-Halwagi, M.M., Gabriel, F. and Harell, D. (2003). Rigorous graphical targeting for resource conservation via material recycle / reuse networks, *Ind. Eng. Chem. Res.*, 42, 4319–4328. DOI: 10.1021/ie030318a.

Hallale, N. (2002). A new graphical targeting method for water minimization, *Advances in Environmental Research*, 6(3), 377–390. DOI: 10.1016/S1093–0191(01)00116–00112.

Kazantzi, V. and El-Halwagi, M.M., (2005). Targeting material reuse via property integration, *Chemical Engineering Progress*, 101(8), 28–37.

Optimal Water© (2006). Software for Water Integration in Industries. Copyright of Universiti Teknologi Malaysia, Johor, Malaysia, <www.cheme.utm.my/prospect>, Accessed on 14/7/2013.

Polley, G.T. and Polley, H.L. (2000). Design better water networks, *Chemical Engineering Progress*, 96(2), 47–52.

Wan Alwi, S.R. and Manan, Z.A. (2008). Generic graphical technique for simultaneous targeting and design of water networks, *Ind. Eng. Chem. Res.*, 47(8), 2762–2777. DOI: 10.1021/ie071487o.

8 Design of Cost-Effective Minimum Water Network (CEMWN)

8.1 Introduction

It is important to note that the concept of Maximum Water Recovery (MWR) only relates to maximum reuse, recycling and regeneration (partial treatment before reuse) of spent water. To achieve the minimum water target, all conceivable methods to reduce water usage through elimination, reduction, reuse/recycling, outsourcing and regeneration need to be considered (Manan and Wan Alwi, 2006). Regenerating wastewater without considering the possibility of elimination and reduction may lead to unnecessary treatment units. The use of water minimisation strategies beyond recycling was first introduced by El-Halwagi (1997) who proposed a targeting technique involving water elimination, segregation, recycling, interception and sink/source manipulation. Hallale (2002) introduced guidelines for reduction and regeneration based on WPA. However, the piece-meal water minimisation strategies proposed do not consider interactions among the process change options as well as the "knock-on effects" of process modifications on the overall process balances, stream data and the economics. The work of Wan Alwi and Manan (2008) has overcome this limitation. The authors have proposed a new framework by using the Water Management Hierarchy (WMH) as a guide to prioritise process changes and the *Systematic Hierarchical Approach for Resilient Process Screening (SHARPS)* strategies as a new cost-screening technique to select the most cost-optimum solution. In Section 8.1, we began by explaining WMH as a foundation for the holistic framework. This is followed by descriptions of a five-step methodology for designing a Cost-Effective Minimum Water Utilisation Network (CEMWN) in Section 8.2. The stepwise application of CEMWN methodology on a semiconductor plant is demonstrated in Section 8.3.

8.2 Water Management Hierarchy

Fig. 8.1 shows the Water Management Hierarchy (WMH) introduced by Wan Alwi and Manan (2008) consisting of five levels, namely (1) source elimination, (2) source reduction, (3) direct reuse/outsourcing of external water, (4) regeneration, and (5) use of freshwater. Each level represents various water management options. The levels are arranged in order of preference, from the most preferred option at the top of the hierarchy (level 1) to the least preferred at the bottom (level 5). Water minimisation is concerned with the first to the fourth levels of the hierarchy.

Source elimination at the top of the hierarchy is concerned with the complete avoidance of freshwater usage. Sometimes it is possible to eliminate water rather than to reduce, reuse or recycle water. Examples include using alternative cooling media

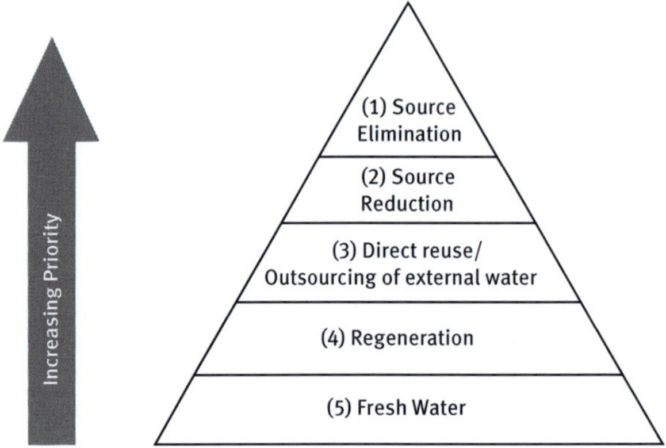

Fig. 8.1: The Water Management Hierarchy (Wan Alwi and Manan, 2008).

such as air instead of water. Even though source elimination is the ultimate goal, often it is not possible to eliminate water completely. One must then try to reduce the amount of water being used at the source of water usage, i.e., certain equipment or processes. Such measures are referred to as *source reduction,* which is the next best option in the WM hierarchy (level 2). Examples of source reduction equipment include water-saving toilet flushing systems and automatic taps. *Source elimination* or *reduction* can be achieved through many ways, e.g. technology and equipment changes, input material changes, good operation practices, product composition changes and reaction changes (Perry and Green, 1998). In the case of good operation practices, it may involve no cost at all with all the other changes the cost varies from low to high.

When it is not possible to eliminate or reduce freshwater at source, wastewater recycling should be considered. Levels 3 and 4 in the WM hierarchy represent two different modes of water recycling – *direct reuse/outsourcing* (level 3) and *regeneration reuse* (level 4). Direct reuse or outsourcing may involve using spent water from within a building or using an available external water source (e.g. rainwater or river water). Through direct reuse (level 3), spent water or external water source is utilised to perform tasks which can accept lower quality water. For example, wastewater from a bathroom wash basin may be directly channelled to a toilet bowl for toilet flushing. Rainwater, on the other hand may be used for tasks which need higher quality water.

Regeneration involves partial or total removal of water impurity through treatment of wastewater to allow the regenerated water to be reused. Examples of regeneration units are gravity settling, filtration, membranes, activated carbon, biological treatment, etc. These regeneration units can be used in isolation or in combination. There are two possible cases of regeneration. Regeneration-recycling involves reuse of treated water in the same equipment or process after treatment. Regeneration-reuse involves reuse of treated water in other equipment after treatment. To increase water availability, the Water Composite Curves and the Pinch concentration can be used to guide regeneration of water sources as follows (Hallale 2002):

1. *Regeneration above the Pinch*: water source(s) in the region above the Pinch are partially treated to upgrade their purity.
2. *Regeneration across the Pinch*: water source(s) in the region below the Pinch are partially treated to achieve purity higher than the Pinch purity.
3. *Regeneration below the Pinch*: water source(s) in the region below the Pinch are partially treated to upgrade their purity. However, the resulting water source is still maintained below the Pinch.

Note that regeneration below and across the Pinch will reduce the freshwater consumption and wastewater generation while regeneration above the Pinch will only reduce wastewater generation. A water regeneration unit can be categorised as fixed outlet concentration (C_{Rout}) and removal ratio (RR) type (Wang and Smith, 1994). For fixed C_{Rout} type, the wastewater concentration outlet is assumed to be the same regardless of the initial inlet concentration (C_{Rin}) of the water source. For RR type, RR value is fixed, C_{Rout} is affected by C_{Rin} and can be calculated by using Equation (8.1), where f_{in} and f_{out} are the inlet and outlet flowrate of the regeneration unit.

$$RR = \frac{f_{in}\,C_{R\,out} - f_{out}\,C_{R\,in}}{f_{in}\,C_{R\,in}} \tag{8.1}$$

Freshwater usage (level 5) should only be considered when wastewater cannot be recycled or when wastewater needs to be diluted to obtain a desired purity. Note that wastewater has to undergo the *end-of-pipe* treatment before discharge to meet the environmental guidelines. Use of freshwater is the least desirable option from the water minimisation point of view and is to be avoided whenever possible. Through the WM hierarchy, the use of freshwater may not be eliminated, but it will become economically legitimate.

8.3 Cost-Effective Minimum Water Network (CEMWN)

The Cost-Effective Minimum Water Network (CEMWN) design procedure is a holistic framework for water management applicable to industry and urban sectors introduced by Wan Alwi and Manan (2008). Fig. 8.2 illustrates five key steps involved in generating the CEMWN, i.e. (1) specify limiting water data, (2) determine the maximum water recovery (MWR) targets, (3) screen process changes using WMH, (4) apply SHARPS strategies and (5) design CEMWN. The first step is to identify the appropriate water sources and water sinks having potential for integration. The next step is to establish the MWR targets the using Water Cascade Analysis (WCA) technique of Manan et al. (2004). The WM hierarchy along with a set of new process screening heuristics is then used to guide process changes to achieve the minimum water targets. The fourth step is to use SHARPS strategies to economically screen inferior process changes.

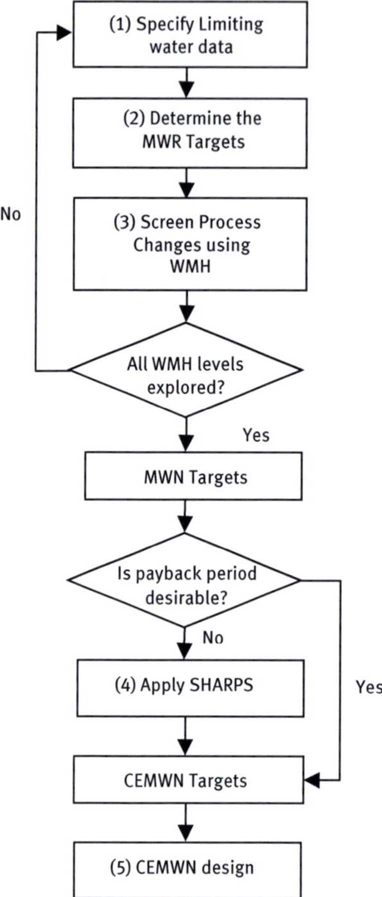

Fig. 8.2: A holistic framework to achieve CEMWN (Wan Alwi and Manan, 2008).

The CEMWN is finally designed using established techniques for design of water networks. The stepwise approach is described in detail next.

Step 1: Specify the limiting water data

The first step is to specify the limiting water data as described in Section 5.5.

Step 2: Determine the MWR targets

The second step is to establish the *base-case* MWR targets, i.e. the overall freshwater requirement and wastewater generation as described in Chapter 6. Note that the *base-case* MWR targets exclude other levels of WMH except reuse and recycling of available water sources and mixing of water sources with freshwater to satisfy water sinks.

Step 3: Screen process changes using WMH

Changes can be made to the flowrates and concentrations of water sources and sinks to reduce the MWR targets and ultimately achieve the MWN benchmark. This was done by observing the basic Pinch rules for process changes and by prioritising as well as assessing all possible process change options according to the WM hierarchy. The fundamental rules to change a process depend on the location of water sources and sinks relative to the Pinch Point of a system:

1. Below the Pinch – Beneficial changes can be achieved by either increasing the flowrate or purity of a source or by decreasing the flowrate or purity requirements of a sink. These changes will increase water surplus below the Pinch thereby reducing the amount of freshwater required.
2. Above the Pinch – There is already a surplus of water above the Pinch, hence any flowrate change made there will not affect the target. An exception to this rule of thumb is for the case where source purity is increased so that it moves to the region below the Pinch as in the case of regeneration.
3. At the Pinch Point – Increasing the flowrate of a source at the Pinch concentration will not reduce the targets.

It is vital to note that implementation of each process change option yields new Pinch Points and MWR targets. In addition, interactions and "knock-on effects" between the process change options should also be carefully considered. It is therefore important that each process change be systematically prioritised and assessed with reference to the revised Pinch Points instead of the original Pinch Point so as to obey the fundamental rules for process changes listed previously and to guarantee that the MWN benchmark is attained. Bearing in mind these constraints, the core of Step 3 is the level-wise hierarchical screening and prioritisation of process change options using the WMH and the following four option-screening heuristics which were sequentially applied to prioritise process changes at each level of WMH. As described below, not all four heuristics are applicable at each level of WMH.

Heuristic 8.1: *Begin process changes at the core of a process.*

Heuristic 8.1 was formulated from the *onion model* for process creation shown in Fig. 8.3 (Smith, 1995). Due to interactions among reaction, separation and recycle, heat and mass exchange network and utility layers, any changes, such as sink elimination should be implemented beginning from the core of a process (reaction system) to the most outer layer (utilities). Excessive water usage at the core of the system causes wastage at the outer layers. Hence improving the core of the system first will eliminate or reduce wastage downstream.

Heuristic 8.1 strictly applied to the process change options at levels 1 and 2 of WMH. Applying Heuristic 8.1 to various source elimination options at level 1 of the WMH will lead to new targets and Pinch Points. For mutually exclusive options, the

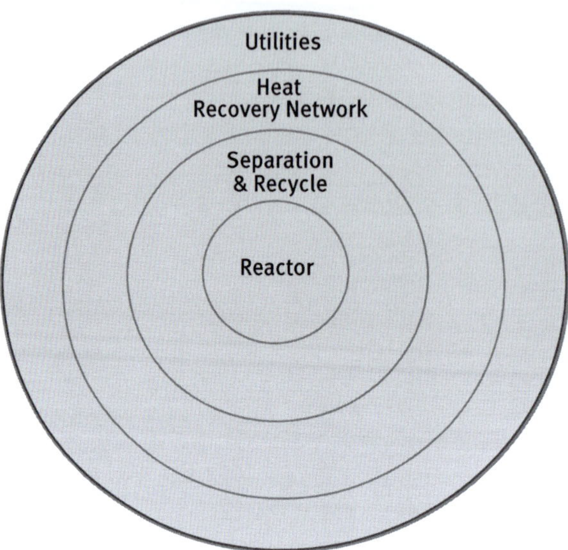

Fig. 8.3: The Onion Diagram as a conceptual model of the hierarchy of the components of a chemical process (Linnhoff et al. 1982).

one giving the lowest revised MWR targets was selected. Heuristic 8.1 was repeated to reduce water at WMH level 2 once all elimination options were explored.

Note also that it is quite common for processes to have independent and non-interacting sources and sinks at various concentrations. For example, reducing water sink for a scrubber in a waste treatment system does not affect the cooling tower sink. In such a case, the sink flowrate above the Pinch (see Rule (i) for process changes mentioned previously) can be reduced using Heuristic 8.2.

> Heuristic 8.2: *Successively reduce all available sinks with a concentration lower than the Pinch point, beginning from the cleanest sink.*

Note that if a dirtier sink were reduced first followed by a cleaner sink, it might be found later that subsequent reduction of a cleaner sink might cause the dirtier sink to lie below the new Pinch Point. Such a situation makes the earlier changes to the dirtier sink meaningless.

If a few sinks exist at the same concentration, it is best to begin by reducing the sink that yields the most flowrate reduction to achieve the biggest savings. Then, proceed to reduce the remaining sinks that exist at concentrations lower than the revised Pinch concentration, as stated in Heuristic 8.3. Heuristics 8.2 and 8.3 were applicable to levels 1 and 2 of the WMH.

> Heuristic 8.3: *Successively reduce the sinks starting from the one giving the biggest flowrate reduction if several sinks exist at the same concentration.*

There exists a maximum limit for adding new water sources (utilities) either obtained externally such as rainwater, river water, snow and borehole water or by regenerating wastewater in order to minimise freshwater in a water distribution system. It is therefore necessary to:

Heuristic 8.4: *Harvest outsourced water or regenerate wastewater only as needed.*

Note that the limit for adding utilities through outsourcing and regeneration corresponds to the minimum utility flowrate (F_{MU}) which leads to minimum freshwater flowrate. To calculate F_{MU}, refer to Section 6.4. Heuristic 8.4 only applies to levels 3 and 4 of the WMH.

The revised MWR targets as well as the four new option-screening heuristics were used as process selection criteria. The screening and selection procedure was hierarchically repeated down the WMH levels to establish the minimum water network (MWN) targets which yielded the maximum scope for water savings. SHARPS strategy was used next to ensure that the savings achieved were cost-effective and affordable.

Step 4: Apply SHARPS strategy

The SHARPS screening technique involves cost estimations associated with water management (WM) options prior to detailed design. It includes a profitability measure in terms of payback period; i.e. the duration for a capital investment to be fully recovered. Note that the payback period calculations for SHARPS as given by Equation (8.2) only concern the economics associated with design of a minimum water network as opposed to the design of an entire plant.

$$Payback\,period\,(yrs) = \frac{Net\,Capital\,Investment\,(\$)}{Net\,Annual\,Savings\,(\$\,/\,y)}. \tag{8.2}$$

Since *SHARPS* is a cost-screening tool, standard plant design *preliminary cost estimation* techniques are used to assess the capital and operating costs of a proposed water system. The equipment, piping and pumping costs built in Equation (8.3) are the three main cost components considered for a building or a plant water recovery system.

$$\Sigma CC = C_{PE} + C_{PEI} + C_{piping} + C_{IC}. \tag{8.3}$$

Where, C_{PE} = Total capital cost for the equipment, \$
               C_{PEI} = Equipment installation cost, \$
               C_{piping} = Water reuse piping cost investment, \$
               $C_{IC,}$ = Instrumentation and controls cost investment, \$

The economics of employing the WM options for grassroots design as well as retrofit cases were evaluated by calculating the net capital investment (NCI) for the MWN using Equations (8.2) and (8.3) as well as the net annual savings (NAS) using Equations (8.4) and (8.5). In the context of SHARPS, the NCI for grassroots refers to the cost difference between the new (substitute) equipment and the base-case equipment. The base-case equipment is the initial equipment used before CEMWN analysis. For the retrofit case, the NCI covers only the newly installed (substitute) system.

$$\textit{Net Capital Investment, \$ (grassroots)} = \Sigma CC_{new\,system} - \Sigma CC_{base\,case}. \qquad (8.4)$$

$$\textit{Net Capital Investment, \$ (retrofit)} = \Sigma CC_{new\,system}. \qquad (8.5)$$

Where, $CC_{new\,system}$ = Capital cost associated with new equipment, \$

$CC_{base\,case}$ = Capital cost for base-case equipment, \$

For example, a new \$300, 6 L toilet flush gives water savings of 6 L per flush as compared to a \$200, 12 L toilet flush (base-case system). For grassroots design, the payback period is therefore based on the NCI given by Equation (8.4), i.e. \$100. For retrofit, the payback period is based on the NCI given by Equation (8.5), i.e. \$300.

The net annual savings (NAS) is the difference between the base-case water operating costs and the water operating costs after employing WM options as in Equation (8.6).

$$NAS = OC_{base-case} - OC_{new}. \qquad (8.6)$$

Where, NAS = Net annual savings (\$/y),

$OC_{base-case}$ = Base-case expenses on water (\$/y), and

OC_{new} = New expenses on water after modifications (\$/y).

The total operating cost of a water system includes freshwater cost, effluent disposal charges, energy cost for water processing and the chemical cost as given by Equation (8.7).

$$OC = C_{FW} + C_{WW} + C_{EOC} + C_c. \qquad (8.7)$$

Where, OC = Total water operating cost

C_{FW} = Costs per unit time for freshwater

C_{FD} = Costs per unit time for wastewater disposal

C_{EOC} = Costs per unit time for energy for water processing

C_c = Costs per unit time for chemicals used by water system.

In order to obtain a cost-effective and affordable water network that achieves the minimum water targets (hereby termed the *Cost-Effective Minimum Water Network* (CEMWN)) within a desired payback period, the new *SHARPS* technique was implemented as follows:

Step 1: Set the desired Total Payback Period (TPP_{set}). The desired payback period can be an investment payback limit set by a plant owner, e.g. two years.

Step 2: Generate an Investment vs. Annual Savings (IAS) composite plot covering all levels of the WM hierarchy. Fig. 8.6 shows a sample of the IAS plot. The gradient of the plot gives the payback period for each process change. The steepest positive gradient (m_4) giving the highest investment per unit of savings represents the most costly scheme. On the other hand, a negative slope (m_3) indicates that the new process modification scheme requires lower investment as compared to the grassroots equipment.

Note that, since most equipment cost is related to equipment capacity through a power law, one is more likely to generate a curved line such as m_5. Hence, in such cases, several data points should be taken to plot a curve for each process change. As in the case of a linear line, a curve moving upward shows that more investment is needed and a curve moving downward shows less investment is needed for an increase in annual savings.

Step 3: Draw a straight line connecting the starting point and the end point of the IAS plot (Fig. 8.4). The gradient of this line is a preliminary cost estimate of the Total Payback Period (TPP) for implementing all options in line with the WM hierarchy. The TPP_{BS} is the total payback period before implementing SHARPS.

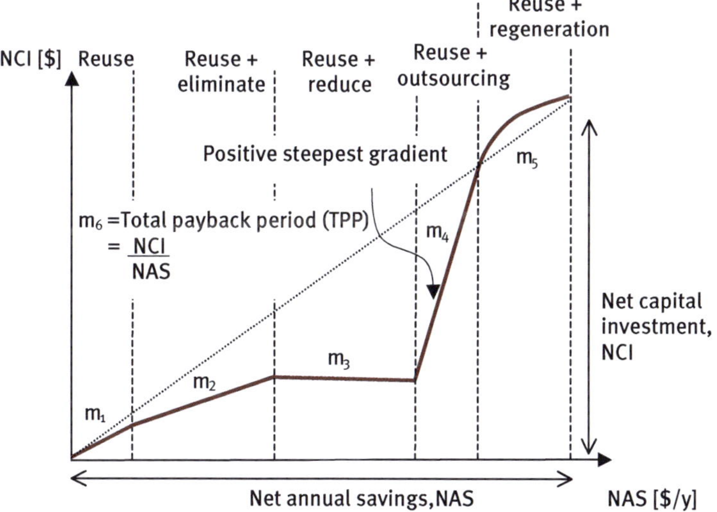

Fig. 8.4: IAS plot covering all levels of WM hierarchy. m_4 is the positive steepest gradient and TPP is the total payback period for a water network (Wan Alwi and Manan, 2008).

Step 4: Compare the TPP_{BS} with the TPP_{set} (the desired total payback period set by a designer).

The total payback period (TPP_{BS}) should match the maximum desired payback period set (TPP_{set}) by a designer. Thus, it is possible to tailor the minimum water network as per the requirement of a plant/building owner.

If $TPP_{BS} \le TPP_{set}$, proceed with network design.

If $TPP_{BS} > TPP_{set}$, two strategies may be implemented.

Strategy 1 – Substitution: This strategy involved replacing the equipment/process that resulted in the steepest positive gradient with an equipment/process that gave a less steep gradient. Note that this strategy does not apply to the reuse line since there is no equipment to replace. To initialise the composite plot, the option that gave the highest total annual water savings should be used regardless of the total investment needed. Hence, to reduce the steepest gradient according to Strategy 1, the process change option giving the next highest total annual saving but with lesser total investment was selected to substitute the initial process option and trim the steepest gradient. Fig. 8.5 shows that substituting the option causing the steepest positive gradient (m_4) with an option that gives a less steep gradient (m_4') yields a smaller TPP value. For example, a separation toilet may be changed to a much cheaper dual-flush toilet that uses a bit more water. TPP$_{AS}$ is the TPP after implementing SHARPS strategies.

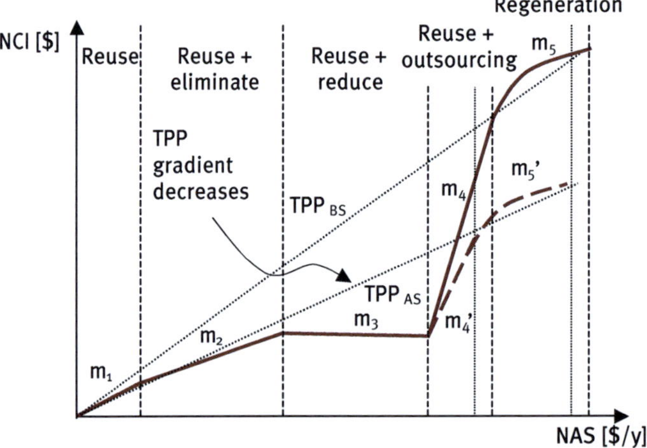

Fig. 8.5: IAS plot showing the revised total payback period when the magnitude of the steepest gradient is reduced using *SHARPS* substitution strategy (Wan Alwi and Manan, 2008).

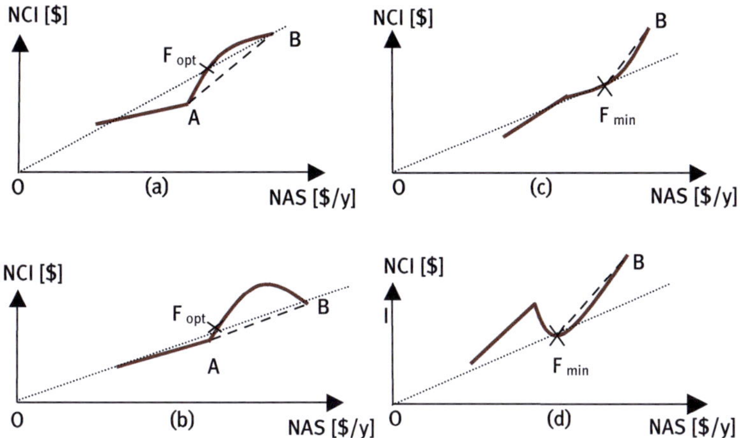

Fig. 8.6: Linearisation of concave curves moving upwards (a) without peak (b) with peak. Convex curves moving upwards linearisation (c) without valley (d) with valley (Wan Alwi and Manan, 2008).

In the case of a curvature, linearisation is necessary to determine the line of steepest gradient. For a projecting concave or convex curve, the linearisation of a curve moving upwards is as follows:

1. Concave curves
   - Connect a straight line to the start (point A) and end (point B) points of the concave curve to obtain a positive gradient (line AB in Figs. 8.6(a) and 8.6(b)). Connect a line from the graph origin (point O) to the end point (point B) of the concave curve. The point where line OB intersects the concave curve is the F_{opt} point. To have beneficial TPP reduction, the concave curve must be reduced below point F_{opt} (for Strategy 2).

2. Convex curve
   - Connect a line from the graph origin (point O) to the minimum point of the convex curve (F_{min}). Connect a line from F_{min} to the end point (point B) of the convex curve to obtain a positive gradient (line F_{min}-B in Figs. 8.6(c) and 8.6(d)). Do not reduce further the line on the left-hand side of F_{min} since this will increase TPP (for Strategy 2).

When the linearised line is the steepest positive gradient, *Strategy 1* is implemented to yield a linearised line with a smaller gradient. Note that the proposed linearisation is only a preliminary guide to screen the most cost-effective option that satisfies a preset payback period.

Strategy 2 – Intensification: The second strategy involves reducing the length of the steepest positive gradient until *TPP*$_{AS}$ is equal to *TPP*$_{set}$. This second strategy is also not applicable for the reuse line since there is no equipment to replace. Fig. 8.7

shows that when the length of the steepest positive gradient (m_4) is reduced, the new gradient line (m_4') gives a less steep gradient, and hence, a smaller TPP. This means that instead of completely applying each process change, one can consider eliminating or partially applying the process change that gives the steepest positive gradient, and hence, a small annual saving compared to the amount of investment. For example, instead of changing all normal water taps to infrared-type, only 50 % of the water taps are changed. If TPP_{AS} is still more than the TPP_{set} even after adjusting the steepest gradient, the length for the next steepest gradient is reduced until TPP is equal to TPP_{set}.

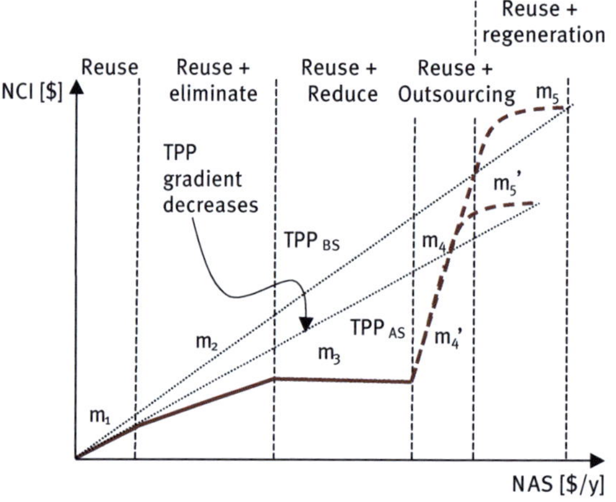

Fig. 8.7: IAS plot showing the revised total payback period with a shorter steepest gradient curve (Wan Alwi and Manan, 2008).

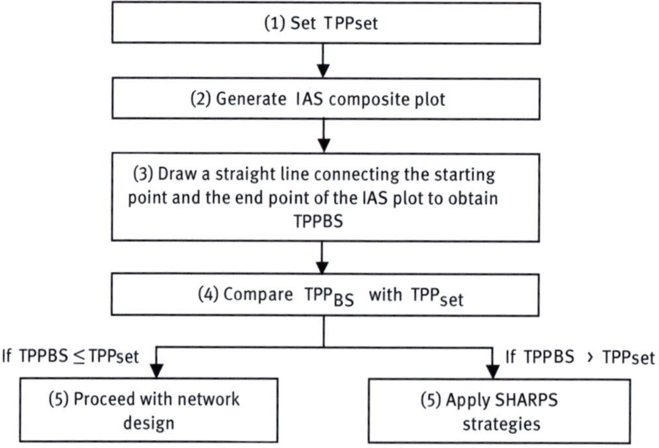

Fig. 8.8: The overall *SHARPS* procedure (Wan Alwi and Manan, 2008).

Similarly, for the case of a projecting concave and convex curves moving upward, it is desirable to reduce the length of the curve until TPP_{set} is achieved, if linearisation of the curve gives the steepest gradient. Both Strategies *1* and *2* should be tested or applied together to yield the best savings. The overall procedure for *SHARPS* is summarised in Fig. 8.8.

Step 5: Network design

Once the CEMWN targets have been established, the next step is to design the Cost-Effective Minimum Water Network (CEMWN) to achieve the CEMWN targets. The water network can be designed using one of the established techniques as described in Chapter 7.

8.4 Industrial case study – a semiconductor plant

The CEMWN method has been implemented to a semiconductor plant in Malaysia by Wan Alwi and Manan (2008). The plant is known as MySEM. MySem uses extensive amounts of water to produce wafer. A large portion of water is used to produce DI or ultrapure water. Creating DI water is among the most expensive and energy-intensive steps in the fabrication process, so any decreases in water sink can result in significant savings. For MySem, creating DI water cost approximately €2.40/m³. Currently MySem is paying €15,952/month (34,618 m³/month) to produce 118 unit of wafer. This amounts to 293 m³ of water needed for each wafer. We will demonstrate the step-by-step procedure to obtain the CEMWN solution by using this case study.

Step 1: Specify the limiting water data

This step involved detailed process survey and line-tracing, establishing process stream material balances and conducting water quality tests. Stream flowrates were either extracted from plant distributed control system (DCS) data or from online data-logging using an ultrasonic flow meter. Depending on the stream audited, tests for total suspended solids (TSS), biological oxygen demand (BOD), chemical oxygen demand (COD), and total dissolved solids (TDS) were made on-site. MySem processes comprised entirely of ultra-pure water and hence TSS were found to be very negligible. BOD was eliminated since there were no biological contaminations. COD was a component of TDS. TDS was ultimately chosen as the dominant water quality parameter for MySem. TDS was monitored using a conductivity meter. Some of the key constraints considered included:

- Water streams with hydrogen fluoride (HF), isopropyl butanol (IPA) and dangerous solvents were not considered as water source.
- Multimedia filter (MMF) backwash was not considered as water source since it contained high TSS.

- Wet bench WB202 and 203 cooling were not reused since they involved acid spillage.
- Black water, i.e. toilet pipes, toilet flushing and office cleaning wastewater were not reused.
- Grey water could only be reused for processes which did not involve body contact.

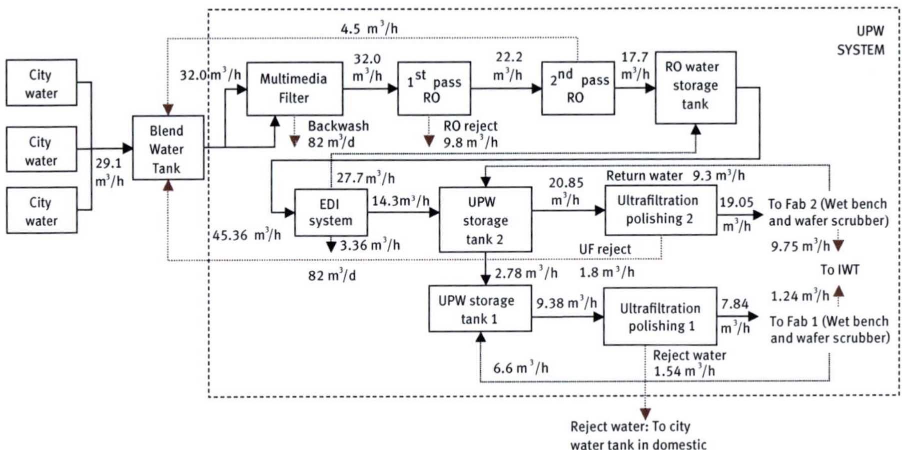

Fig. 8.9: MySem DI water balance.

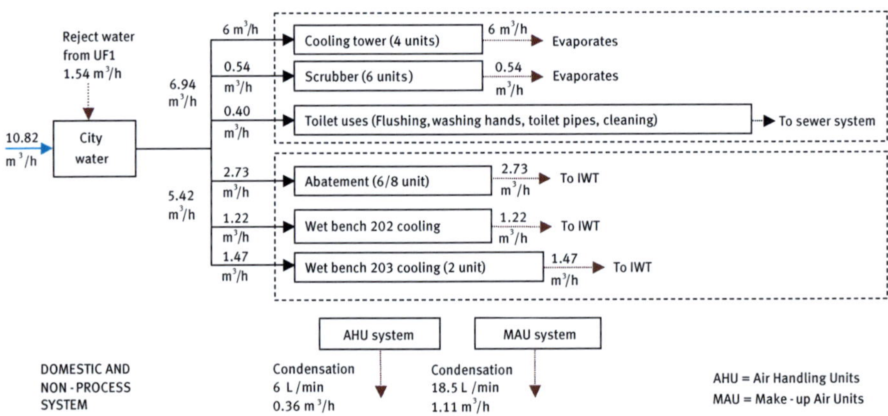

Fig. 8.10: MySem non-process water balance (October-November).

Portable ultrasonic flow meters and MySem online monitoring systems were used to establish water balances. Figs. 8.9 and 8.10 show the water-using processes in terms of total flowrate in m³/h for MySem. The Process Flow Diagram of MySem was divided into three sections including (i) DI water balance, (ii) domestic water balance and (iii) fabrications (Fab) water balances.

The relevant water streams having potential for recycling were extracted into a table of *limiting water data* comprising process sources and sinks. Table 8.1 shows the water sources and sinks extracted for MySem listed in terms of flowrate and contaminant concentration. The contaminant (C, in ppm) in Table 8.1 represents TDS.

Tab. 8.1: Limiting water data for MySem.

	Sink	F, t/h	C, ppm		Source	F, t/h	C, ppm
SK1	MMF inlet	32.0	52	SR1	MMF rinse	1.33	48.0
SK2	Cooling tower	6.00	100	SR2	RO reject 1st pass	9.80	70.4
SK3	Abatement	2.73	100	SR3	EDI reject	3.36	48.6
SK4	Scrubber	0.54	100	SR4	WB101 rinse water, idle	0.38	0
SK5	Toilet flushing	0.08	100	SR5	WB101 rinse water, operation	0.07	4608
SK6	Wash basin	0.01	52	SR6	WB102 rinse water, idle	0.22	0
SK7	Ablution	0.15	52	SR7	WB102 rinse water, operation	0.07	4480
SK8	Toilet pipes	0.12	52	SR8	WB201 rinse water, idle	0.76	0
SK9	Office cleaning	0.05	52	SR9	WB201 rinse water, operation	0.03	23360
SK10	MMF backwash	2.08	52	SR10	WB202 rinse water, idle	3.48	0
SK11	MMF rinse	1.33	52	SR11	WB202 rinse water, operation	0.07	163.2
SK12	WB203 cooling	1.47	52	SR12	WB203 rinse water, idle	3.63	0
SK13	WB202 cooling	1.22	52	SR13	WB203 rinse water, operation	0.28	928
				SR14	MAU	1.11	6.4
				SR15	AHU	0.36	11.5
				SR16	Cassette cleaner	0.08	0
				SR17	Abatement	2.73	105.6
				SR18	Wafer scrubber	0.54	12.8
				SR19	RO reject 2nd pass	4.50	19.2
				SR20	UF1 reject	1.54	19.2
				SR21	UF2 reject	1.80	0
				SR22	Heater WB101	0.46	0
				SR23	Wash basin	0.01	60
				SR24	Ablution	0.15	40

Step 2: Determine the maximum water recovery (MWR) targets

This step involved establishing the base-case MWR targets using WCA technique that was incorporated in Water MATRIX© (2006) software developed in Universiti Teknologi Malaysia (UTM). Table 8.2 is the water cascade table (WCT) generated by Water MATRIX© (2006) for MySem showing the freshwater and wastewater flowrate targets at $F_{FW} = 11.04$ t/h and $F_{IWT} = 0.02$ t/h. Note from Table 8.2 that the cleanest water targeted at 0 ppm concentration actually referred to DI water (F_{DI}) to be supplied to the blend water tank instead of freshwater. This was because freshwater for MySem had a concentration of 30 ppm. The source water flowrate at 30 ppm shown in Table 8.2 was actually the amount of freshwater supply needed. Water MATRIX© computed $F_{DI} = 0$ t/h and $F_{FW} = 11.04$ t/h.

The F_{IWT} of 0.02 t/h shown in Table 8.2 only comprised the IWT wastewater that was considered for reuse. The total IWT should also include 5.69 t/h IWT wastewater not available for reuse due to chemical contamination, giving a total IWT of 5.71 t/h to be discharged for MySem as shown in Table 8.3. Considering the present integration schemes implemented by MySem, the initial FW and total IWT flowrates were at 39.94 t/h and 34.45 t/h. Equations (8.8) and (8.9) gave FW and IWT reduction of 72.4 % and 83.4 %.

$$\text{FW savings}, \% = \frac{\text{FW flowrate before WPA} - \text{FW flowrate targets after WPA}}{\text{FW flowrate before WPA}} \times 100. \tag{8.8}$$

$$\text{IWT savings}, \% = \frac{\text{IWT flowrate before WPA} - \text{Total IWT to be discharged}}{\text{IWT flowrate before WPA}} \times 100. \tag{8.9}$$

Tab. 8.2: Base-case maximum water recovery targets for MySem (without process changes).

C_k, ppm	ΔC_k, ppm	ΣF_{Ski}, t/h	ΣF_{SRj}, t/h	$\Sigma F_{SKi} + \Sigma F_{SRj}$, t/h	F_C, t/h	Δm, kg/h	Cum. Δm, kg/h	$F_{FW, cum}$, t/h	F_C, t/h
					0.00				$F_{DI} = 0.00$
0			10.81	10.81			0.00		
	6.40				10.81	69.17			10.81
6.4			1.11	1.11			69.17	10.81	
	5.12				11.92	61.02			11.92
11.52			0.36	0.36			130.19	11.30	
	1.28				12.28	15.72			12.28
12.8			0.54	0.54			145.91	11.40	
	6.40				12.82	82.04			12.82

Tab. 8.2: Base-case maximum water recovery targets for MySem (without process changes).

C_k, ppm	ΔC_k, ppm	ΣF_{Ski}, t/h	ΣF_{SRj}, t/h	$\Sigma F_{SKi} + \Sigma F_{SRj}$, t/h	F_C, t/h	Δm, kg/h	Cum. Δm, kg/h	$F_{FW, cum}$, t/h	F_C, t/h
19.2			6.04	6.04			227.94	11.87	
	10.80				18.86	203.67			18.86
30			11.04	$F_{FW} = 11.04$			431.61	14.39	
	10.00				29.90	298.98			29.90
40			0.15	0.15			730.59	18.26	
	8.00				30.05	240.38			30.05
48			1.33	1.33			970.97	20.23	
	0.64				31.38	20.08			31.38
48.64			3.36	3.36			991.05	20.38	
	3.36				34.74	116.72			34.74
52		−38.43	0.00	−38.43			1107.77	21.30	
	8.00				−3.69	−29.54			−3.69
60			0.01	0.01			1078.24	17.97	
	10.40				−3.68	−38.29			−3.68
70.4			9.80	9.80			1039.95	14.77	
	29.60				6.12	181.09			6.12
100		−9.35		−9.35			1221.04	12.21	
	5.60				−3.23	−18.10			−3.23
105.6			2.73	2.73			1202.94	11.39	
	58.40				−0.50	−29.32			−0.50
164			0.07	0.07			1173.62	7.16	
	764.00				−0.43	−330.81			−0.43
928			0.28	0.28			842.81	0.91	
	3552.00				−0.16	−550.56			−0.15
4480			0.07	0.07			292.25	0.07	
	128.00				−0.09	−11.01			−0.09
4608			0.07	0.07			281.24	0.06	
	18752.00				−0.02	−281.28			−0.01
23360			0.03	0.03			−0.04	**0.00**	(Pinch)
					0.02	0.00			$F_{IWT} = 0.02$

Tab. 8.3: Amount of IWT and domestic wastewater before and after integration.

Utility	Before MWR (t/h)	After MWR (t/h)	% reduction
Total freshwater	39.94	11.04	72.4
Total IWT wastewater	34.45	5.71	83.4
Total domestic WW	0.41	0.25	39.0

Step 3: WMH-guided screening and selection of process options

After calculating the base-case MWR targets, all potential process changes to improve the MySem water system were listed according to the various WMH levels as shown in Table 8.4. Central to the MWN approach is the level-wise hierarchical screening and prioritisation of process change options using the water management hierarchy (WMH) and three new option-screening heuristics which were sequentially applied to prioritise process changes. MySem had initially selected process changes based on the net annual savings (NAS) associated with each process change as shown with check marks in column 3 (MySem, 2005) of Table 8.4. The ultimate minimum water targets obtained after MWN analysis were given check marks in column 4 of Table 8.4. The steps for screening the options according to the WMH are described next.

1. Source elimination

The Pinch Point obtained from the base-case MWR targeting stage was at 23,360 ppm (see the water cascade table, i.e., Table 8.2). In order to maximise freshwater savings, the top priority was to consider eliminating water sinks above the Pinch Point in the cascade table, i.e. any sink with concentration less than 23,360 ppm. Note that all the water sinks in Table 8.1 met this criterion. All possible means to change processes or the existing equipment to new equipment to eliminate water sinks were considered. From Table 8.4 it was possible to eliminate SK12 and SK13 by changing wet bench 202 and 203 quartz tanks, that initially needed continuous water for cooling, to Teflon tanks. Not only did this option eliminate water requirement, it also avoided tank cracking as a result of a sudden temperature drop. Elimination of SK12 and SK13 resulted in new water targets at 8.3525 t/h freshwater and 0.0215 t/h IWT (see third row of Fig. 8.11). The Pinch Point was maintained at 23,360 ppm.

2. Source reduction

After eliminating SK12 and SK13, the next process change considered according to the WM hierarchy was to reduce sinks above the Pinch Point in the cascade table; i.e. any sink with concentration lower than 23,360 ppm. Table 8.4 lists a few process change options related to source reduction. Following Heuristic 8.2, the sink at the lowest contaminant concentration (52 ppm) was reduced first. There follow possible source reduction process changes (listed in Table 8.4) affecting sinks SK1, SK10 and SK11 (all

Tab. 8.4: Various process change options applicable for MySem.

WMH	Strategy	Option selected based on NAS	Option selected based on MWN procedure
Elimination	Abatement		
	Option 2 (decommissioning)	X	X
	WB 202 and 203 cooling	✓	✓
Reduction	WB reduction in Fab 1 and 2	✓	✓
	Heater reduction	✓	✓
	Fab 1 return reduction	✓	✓
	Abatement		
	Option 1 (0.5gpm during idle)	X	X
	Option 3 (recirculation)	✓	X
	Option 4 (on demand)	X	✓
	Option 5 (pH analysis)	X	X
	Increase RO system recovery/ install 3rd stage	✓	✓
	EDI return reduction		
	Option 1 (decommissioning)	X	✓
	Option 2 (run intermittent)	✓	X
	Domestic reduction	✓	X
	Cooling tower reduction using N2	✓	✓
	MMF reduction by NTU analysis	✓	✓
Reuse	Total reuse	✓	✓
Outsourcing	RW harvesting	✓	✓
Regeneration	Treat all WB water	X	✓

(✓) for selected option, (X) for eliminated option by MySem.

located at 52 ppm) and sources SR1 to SR13 and SR19 to SR22 simultaneously due to the interactions between equipment in the water system of the DI plant:
– Wet bench flowrate reduction to the minimum during idle mode,
– Recirculating hot water and switching heater on sink for heater WB201,
– Reduction of Fab 1 return flowrate by changing to variable speed pump,
– Decommissioning three EDI units instead of running four units. Note that a sharp decrease in flowrate due to upstream process changes made it possible to reduce

by three EDI units (Option 2 from Table 8.4 was rejected due to the increase in FW and IWT targets. Option 1, i.e. decommissioning three EDI units, reduced the FW and IWT targets to 6.3038 t/h and 0.0378 t/h (Table 8.5) and was hence implemented),

- Increase rate of recovery for reverse osmosis system,
- Decrease multimedia filter backwash and rinsing time.

The FW and IWT after application of each process change are summarised in Fig. 8.11.

WMH levels	Specific process changes	New FW target, t/h	New IWT+WW [based on limiting data] target, t/h	New Pinch Point concentration, ppm	New total IWT (all IWT considered) target, t/h
Initial	None	39.94	34.85	22360	34.450
Reuse	Base case	11.040	0.0190	22360	5.7090
Elimination	Eliminate WB cooling 202 (SK 13) and 203 (SK 12)	8.3525	0.0215	22360	3.0215
Reduction	WB reduction in Fab 1 and Fab 2	6.7518	0.0258	4608	1.4216
	Heater WB201	6.7314	0.0264	4608	1.4454
	Fab 1 return	6.6094	0.0354	4608	1.3109
	Option 1: EDI decommissioning	6.3038	0.0378	4608	1.0112
	Increase RO rate of recovery Reduce multimedia filter	6.2110	0.0380	4608	0.9132
	backwash and rinsing time	6.0857	0.0387	4608	0.7879
	Option 4: Reduce abatement pollution system (SK3 = 0.57 t/h and SR 17 = 0.57 t/h).	6.0831	0.0361	4608	0.7853
	Cooling tower reduction (SK 2 = 5.86 t/h)	5.9452	0.0382	4608	0.7874
Outsourcing	Add a source, SR 25=0.11 t/h of C = 16 ppm by harvesting rainwater	5.8349	0.0379	4608	0.7871
Regeneration	Regenerate remaining IWT to the maximum flowrate for a source from to C=52 ppm.	5.7970	0	4608	0.7492
Minimum water network (MWN) targets		5.7970	0	4608	0.7492

Fig. 8.11: The effects of WMH-guided process changes on the maximum water reuse/recovery targets and Pinch location.

Tab. 8.5: Various effects of EDI options on water targets.

EDI system	FW target, t/h	IWT target, t/h
Initial EDI flow rate	6.6094	0.0354
Option 1 (decommission 3 EDI unit)	6.3038	0.0378
Option 2 (run intermittently)	6.9281	0.0351

Using Heuristic 8.1, the source reduction process changes for the DI water system shown in Table 8.4 were implemented from the core of the process (wet bench systems) to the most outer layer (multimedia filter). Excessive water usage at the core of the system was the main reason for water wastage at the outer layers. Hence, improving the core of the system first will reduce wastage downstream. Implementation of the entire range of process changes, from wet bench to MMF (multimedia filter) listed in Table 8.4 gave revised freshwater and wastewater targets at 6.0857 t/h and 0.0387 t/h and a new Pinch concentration at 4,608 ppm (refer to the ninth row of Fig. 8.9). Since there were no other sinks at 52 ppm, following Heuristic 8.2, the sinks with the next lowest contaminant concentration (100 ppm) were considered next. For MySem, SK2, SK3 and SK6 existed at the same concentration of 100 ppm. SK3 which yielded the biggest flowrate reduction was chosen first followed by SK2 and SK6 according to Heuristic 8.3.

The pollution abatement system sink (SK3) existed at 100 ppm. Initially, the pollution abatement system sinked 2.73 t/h of water (SK3) and produced 2.73 t/h of IWT (SR17). Table 8.6 shows five possible options to reduce the abatement system sink. Option 3, which was predicted to yield the highest savings, was initially chosen by MySem prior to the MWN approach (Table 8.6 column 3). However, as shown in Table 8.6, option 3 actually increased the freshwater target by 2.2% to 6.2183 t/h. This was because the introduction of a recirculation system that produces no wastewater but relied on the makeup water sink (option 3) reduced the amount of wastewater that

Tab. 8.6: Effects of abatement options on water targets.

Abatement system	F_D, t/h	F_S, t/h	FW target, t/h	IWT target, t/h
Initial abatement flow rate	2.73	2.73	6.0857	0.0387
Option 1 (0.5gpm during idle)	1.11	1.11	6.0837	0.0367
Option 2 (decommissioning)	1.36	1.36	6.0840	0.0370
Option 3 (recirculation)	0.14	0.00	6.2183	0.0333
Option 4 (on demand)	0.57	0.57	6.0831	0.0361
Option 5 (pH analysis)	0.79	0.79	6.0833	0.0363

could potentially be reused for MySem as a whole, thereby leading to increased freshwater target. Option 4 in Table 8.6 gave the highest freshwater and IWT savings. Choosing option 4 led to new freshwater and IWT targets at 6.0831 t/h and 0.0361 t/h with Pinch Point maintained at 4,608 ppm (refer to the tenth row of Fig. 8.9).It was also possible to reduce sink SK2 which also existed at 100 ppm. SK2, which was the cooling tower makeup, had the second highest flowrate reduction. Heat Exchange between the cooling tower circuit and the liquid nitrogen circuit had the potential to reduce SK2 to 5.86 t/h, and ultimately the water targets to 5.9452 t/h freshwater and 0.0382 t/h wastewater (see eleventh row of Fig. 8.9). The Pinch Point was maintained at 4,608 ppm.

Sink SK6 (wash basin) and SK7 (ablution) were reduced to 0.002 t/h and 0.035 t/h respectively by changing the normal water taps to laminar taps. This also reduced sources SR23 and SR24. However, when targeted using Water-MATRIX$^©$, the freshwater and wastewater targets increased slightly to 5.9455t/h and 0.0385 t/h. Hence this process change was rejected.

3. External water sources

The next process change according to the WM hierarchy was to add external water sources at a concentration lower than the new Pinch Point concentration of 4,608 ppm. Based on MySem's available roof area and the rain distribution, it was possible to harvest 0.11 t/h (maximum design limit, $F_{max\ design}$) of rainwater at concentration of 16 ppm as a new water source, SR25. This option had potential to reduce the freshwater and IWT targets to 5.8349 t/h and 0.0379 t/h respectively (refer to the 12th row of Fig. 8.11). The Pinch Point was maintained at 4,608 ppm.

4. Regeneration

Regeneration was the final process change considered according to the WM hierarchy. Freshwater savings could only be realised through regeneration above or across the Pinch. Regenerating all 'WB201 in-operation' (SR9) at 23,360 ppm and 0.0201 t/h (maximum utility flowrate, F_{MU}, obtained using trial and error method in WCA) of 'WB101 in-operation' (SR5) at 4,608 ppm to 52 ppm (the regenerated sources were named SR26) by carbon bed, EDI and ultraviolet (UV) treatment systems reduced the freshwater and IWT targets to 5.797 t/h and 0 t/h respectively (Table 8.7). Considering the IWT excluded from integration, the new IWT flowrate after regeneration was 0.7492 t/h. This corresponded to 85.5 % freshwater and 97.8 % industrial wastewater reductions. The Pinch Point was maintained at 4,608 ppm.

The *minimum water targets* were ultimately obtained after considering all options for process changes according to the WM hierarchy. Note that targeting the maximum water recovery only through reuse and regeneration resulted in savings of up to 72.4 % freshwater and 83.4 % wastewater for MySem. Instead, following the holistic framework guided by the WM hierarchy enabled potential freshwater and wastewater reductions of up to 85.5 % and 97.8 %, towards achieving the minimum water network (MWN) design.

8.4 Industrial case study – a semiconductor plant ⸺ **219**

Tab. 8.7: MySem water targets after implementation of MWN technique.

C_k, ppm	ΔC_k, ppm	ΣF_{Ski}, t/h	ΣF_{SRj}, t/h	ΣF_{SKi} + ΣFSR, t/hj	F_C, t/h	Δm, kg/h	Cum. Δm, kg/h	$F_{FW,cum}$, t/h	F_C, t/h
					0.00				F_{DI} = 0.00
0		1.727	1.73				0.00		
	6.40				1.73	11.05			1.73
6.4		1.11	1.11				11.05	1.73	
	5.12				2.84	14.53			2.84
11.52		0.36	0.36				25.58	2.22	
	1.28				3.20	4.09			3.20
12.8		0.54	0.54				29.67	2.32	
	3.20				3.74	11.96			3.74
16		0.11	0.11				41.63	2.60	
	3.20				3.85	12.31			3.85
19.2		0.776	0.78				53.94	2.81	
	10.80				4.62	49.93			4.62
30		F_{FW} = 5.797	5.80				103.87	3.46	
	10.00				10.42	104.20			10.42
40		0.15	0.15				208.07	5.20	
	8.00				10.57	84.56			10.57
48		0.169	0.17				292.63	6.10	
	0.64				10.74	6.87			10.74
48.64		0.84	0.84				299.50	6.16	
	3.36				11.58	38.91			11.58
52		−6.528	0.0381	−6.49			338.41	6.51	
	8.00				5.09	40.71			5.09
60		0.01	0.01				379.12	6.32	
	10.40				5.10	53.03			5.10
70.4		1.153	1.15				432.15	6.14	
	29.60				6.25	185.06			6.25
100		−7.05		−7.05			617.21	6.17	
	5.60				−0.80	−4.47			−0.80
105.6		0.57	0.57				612.74	5.80	

	58.40			−0.23	−13.31			−0.23
164		0.024	0.02			599.43	3.66	
	764.00			−0.20	−155.78			−0.20
928		0.081	0.08			443.65	0.48	
	3552.00			−0.12	−436.54			−0.12
4480		0.069	0.07			7.11	0.00	
	128.00			−0.05	−6.90			−0.05
4608		0.0539	0.05			0.21	0.00	**(Pinch)**
				0.00	0.00			$F_{IWT} = 0.00$

Step 4: Apply SHARPS strategy

The desired payback period (TPP_{set}) was set at 4 month (0.33 y) by MySem management. Fig. 8.10 shows the IAS plot after MWN analysis for the MySem retrofit. Since this is a retrofit case, Equations (8.1), (8.2) and (8.5) mentioned previously were used. The total payback period to attain the MWN targets before *SHARPS* screening was 0.38 y. Since the initial total payback periods were more than the TPP_{set}, i.e. 0.33 y, SHARPS strategies were applied to fulfil the TPP_{set} specified.

Fig. 8.12 shows that *regeneration* process change gives the steepest gradient. Focusing on *Strategy 1*, there was no other option for *regeneration* process change. Hence, *Strategy 1* which called for equipment substitution could not be implemented.

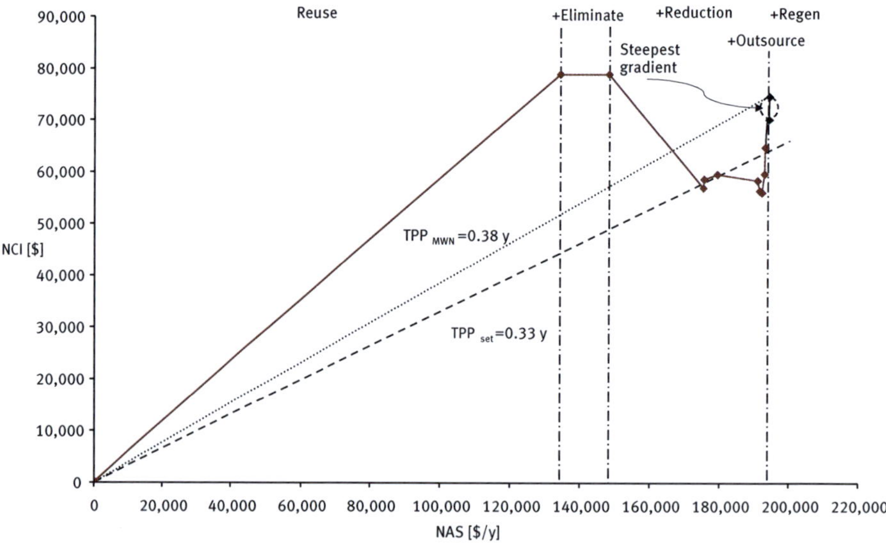

Fig. 8.12: IAS plot for MWN retrofit.

Focusing on *Strategy 2*, when no *regeneration* was applied, the total payback period reduces to 0.36 y as illustrated in Fig. 8.13, which still does not achieve *TPP*$_{set}$. Thus, the next steepest gradient was observed.

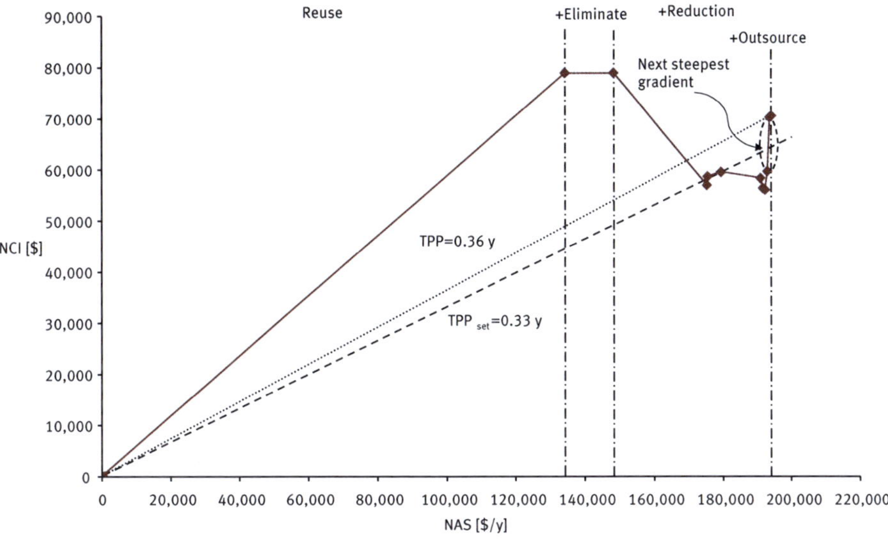

Fig. 8.13: IAS plot after eliminating regeneration curve.

The next steepest gradient was noted to be *cooling tower* process change. Again, *Strategy 1* cannot be applied since only one option exists for that process change. It can be noted that the cooling tower line is a concave curve without peak. F_{opt} is noted to be at the end of the curve as shown in Fig. 8.14. Hence, based on the linearisation rule explained earlier, any reduction of the cooling tower curve will be beneficial. Using *Strategy 2*, reducing the *cooling tower* process change curve by only partially applying nitrogen cooling yielded the final IAS plot shown in Fig. 8.15 that achieved the specified payback of 0.33 y. SK2 was reduced only to 5.966 t/h instead of 5.86 t/h initially. This scheme had reduced 5.94 t/h of freshwater and 0.79 t/h IWT total flowrate.

Hence, the application of *SHARPS* screening has successively achieved the *TPP*$_{set}$ of 4 months with 85.1 % and 97.7 % of freshwater and wastewater reduction prior to design. This is the final cost CEMWN target for MySem. The effect of applying each scheme on freshwater and IWT are illustrated in Fig. 8.16. Using the pre-design cost estimate method, the system needs approximately a net total investment of $64,500 and will give a net annual savings of $193,550.

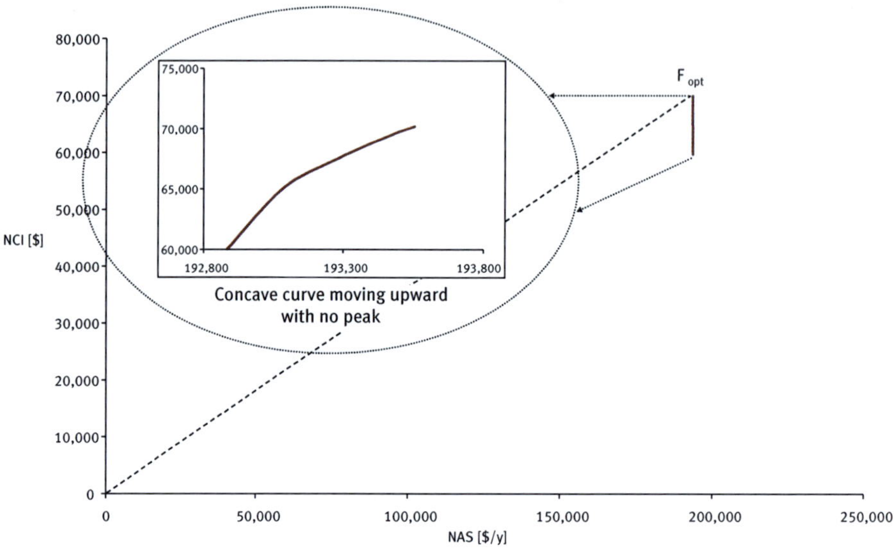

Fig. 8.14: F_{opt} for cooling tower concave curve moving upwards (without peak).

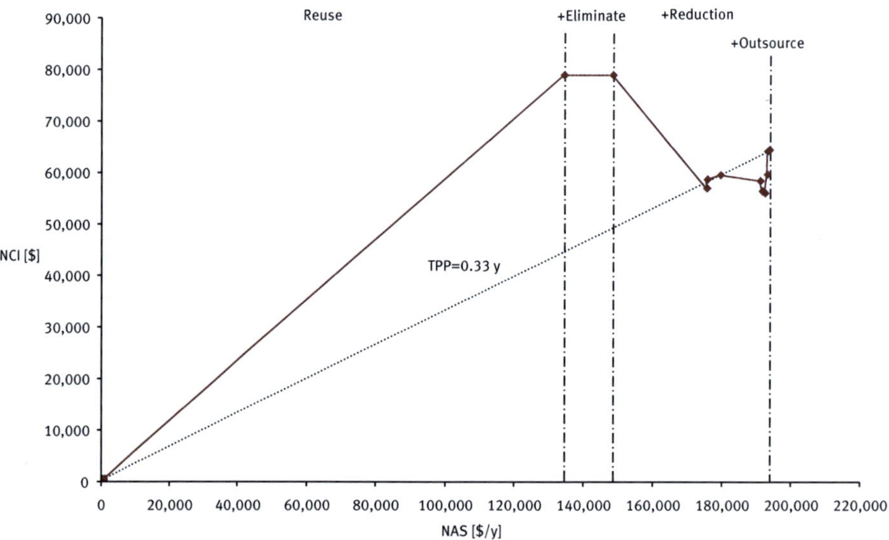

Fig. 8.15: Final IAS plot after *SHARPS* analysis.

WMH levels	Specific process changes considered	New FW target, t/h	New IWT+WW [based on limiting data] target, t/h	New Pinch Point concentration, ppm	New total IWT (all IWT considered) target, t/h
Initial	None	39.94	34.85	22,360	34.4500
Reuse	Base case	11.0400	0.0190	22,360	5.7090
Elimination	Eliminate WB cooling 202 (SK 13) and 203 (SK 12)	8.3525	0.0215	22,360	3.0215
Reduction	WB reduction in Fab 1 and Fab 2	6.7518	0.0258	4,608	1.4216
	Heater WB201 reduction	6.7314	0.0264	4,608	1.4454
	Fab 1 return reduction	6.6094	0.0354	4,608	1.3109
	Option 1: EDI decommissioning	6.3038	0.0378	4,608	1.0112
	Increase RO rate of recovery	6.2110	0.0380	4,608	0.9132
	Reduce multimedia filter backwash and rinsing time	6.0857	0.0387	4,608	0.7879
	Option 4: Reduce abatement pollution system (SK 3 = 0.57 t/h and SR 17 = 0.57 t/h).	6.0831	0.0361	4,608	0.7853
	Cooling tower reduction (SK 2 = 5.966 t/h)	6.0496	0.0366	4,608	0.7858
Outsourcing	Add a source, S25=0.11 t/h of C = 16 ppm by harvesting rainwater	5.9392	0.0362	4,608	0.7854
CEMWN targets		5.9392	0.0362	4608	0.7854

Fig. 8.16: Final CEMWN targets after SHARPS analysis.

Step 5: Network design

Network design based on Source and Sink Allocation Curves (SSAC) methodology proposed in Chapter 7 was used. The Source and Sink Composite Curves (SSCC) and Source and Sink Allocation Curves (SSAC) for the MySem retrofit system are shown in Figs. 8.17 and 8.18. It can be seen that the SSAC are very complicated with many streams. By directing the reuse wastewater to existing tanks, the system was divided into three categories: domestic city water tank (for sinks SK6 to SK9), process city water tank (for sinks SK2 to SK5) and DI blend water tank (for sinks SK1, SK10 and SK11). The new simplified retrofit SSAC and Network Mapping Diagram are as shown in Fig. 8.19. The final network that achieved the CEMWN target for MySem is shown in Fig. 8.20.

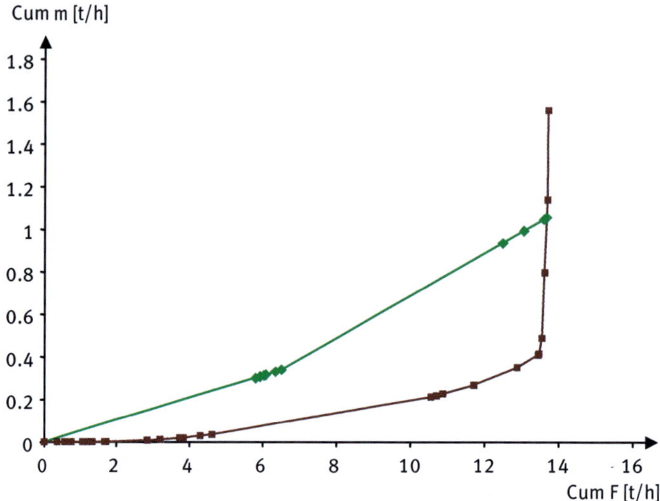

Fig. 8.17: Source and Sink Composite Curves for MySem.

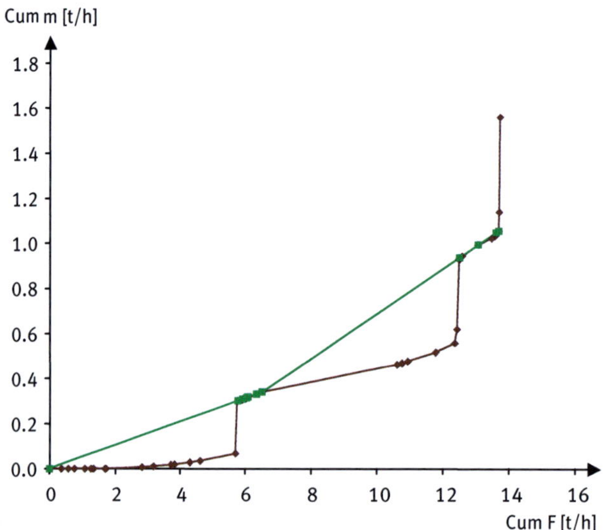

Fig. 8.18: Source and Sink Allocation Curves for MySem.

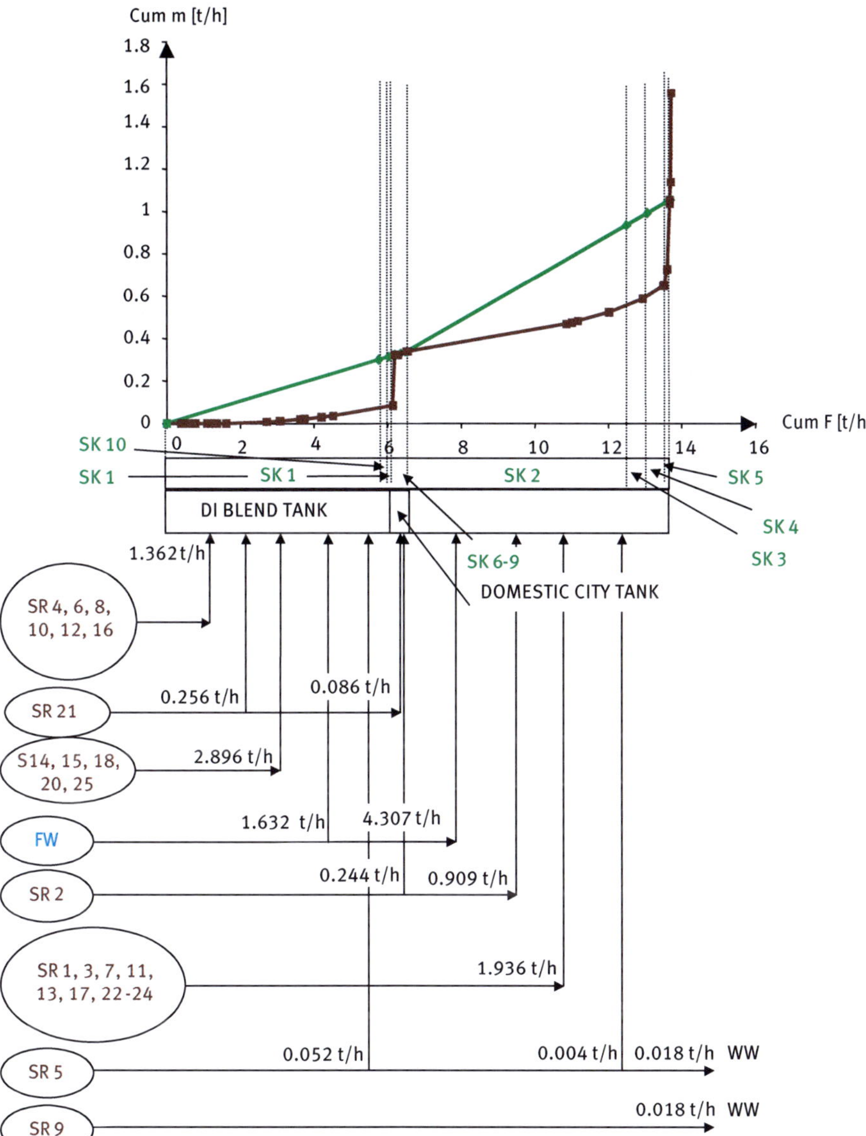

Fig. 8.19: Network Allocation Diagram based on simplified SSAC for MySem retrofit.

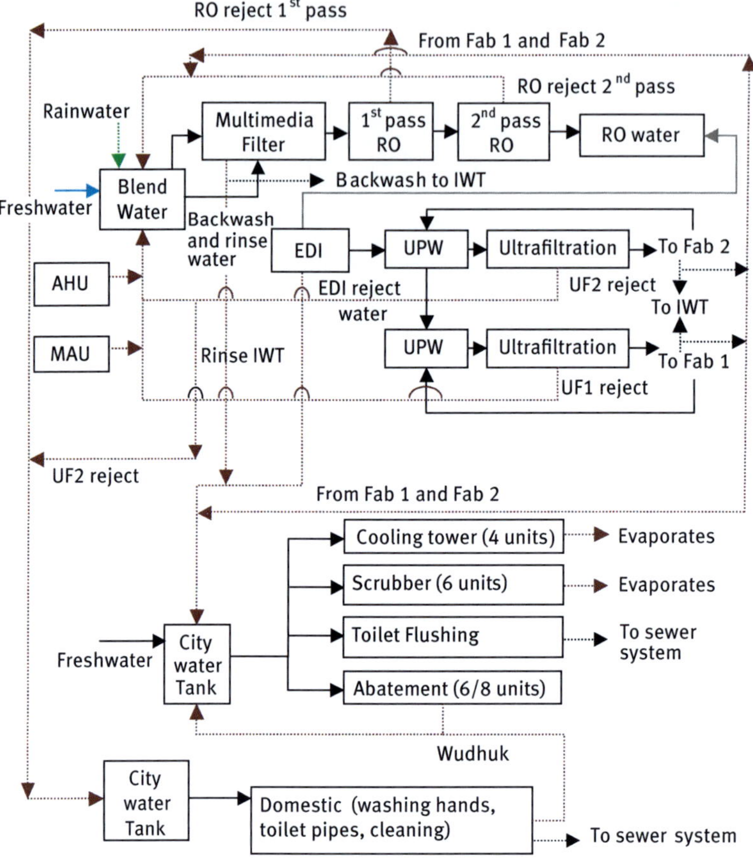

Fig. 8.20: MySem retrofit DI water balance and non-process water balance after CEMWN analysis, achieving 85.5 % freshwater and 97.7 % IWT reductions within 4 months payback.

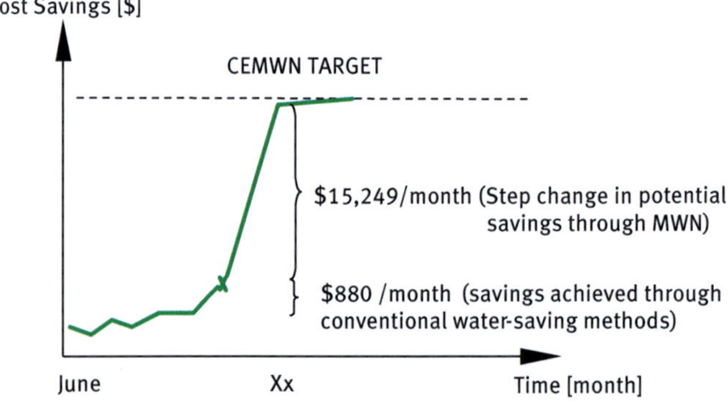

Fig. 8.21: Savings achieved by MySem in comparison to savings predicted through CEMWN technique (Wan Alwi and Manan, 2008).

8.4.1 Using CEMWN targets as reference benchmarks

CEMWN targets can be used for own performance and international water reduction benchmarking guides. For example, the CEMWN target for MySem, which also corresponds to the best achievable benchmark targets, are a target of freshwater flowrate of 5.94 t/h and a total IWT flowrate of 0.79 t/h. This represented 85.1 % freshwater and 97.7 % IWT reduction. Hence, these were the best performance benchmark targets (Fig. 8.21) that MySem needed to achieve. The application of total reuse only using Water Pinch Analysis (WPA) method yielded lower water savings potentials of 72.4 % freshwater and 83.4 % wastewater reduction with a 0.59 y payback period. November 2005 water bills had shown that all the conventional water reductions strategies applied by MySem had only managed to reduce freshwater usage from 42.6 t/h to 40.24 t/h representing a saving of \$880/month. An estimated total saving of 193,550/y was predicted with the implementation of the CEMWN method. A preliminary cost estimate indicated that this best performance required an investment of approximately \$64,507 with a payback period of 0.33 y. Note that the schemes proposed and listed in Fig. 8.16 could also be gradually implemented as part of the company's longer-term utility savings program in line with its quality management practices.

Once the best performance benchmark was established through the CEMWN method, the predicted maximum savings of MySem were then compared with the international benchmark. The International Technology Roadmap for Semiconductors (ITRS 2001) had aimed to reduce high purity water (HPW) consumption from the current rate of 6–8 m^3 in 2005 to 4–6 m^3 per wafer by 2007 (Wu et al., 2004). After CEMWN analysis, MySem had potential to use 4.06 m^3 of DI water per wafer for Fab 1 and 13.73 m^3 of DI water per wafer for Fab 2, down from its previous consumption of 6.3 and 72.4 m^3 of DI water per wafer respectively. Fab 1 had potential to meet the ITRS 2001 target. Fab 2 however was far from this ITRS target due to its wafer production rate of well below the design capacity.

References

El-Halwagi, M.M. (1997). *Pollution prevention through process integration: systematic design tools.* San Diego: Academic Press.

Hallale, N. (2002). A new graphical targeting method for water minimization. *Advances in Environmental Research.* 6 (3), 377–390.

International Technology Roadmap For Semiconductors (ITRS) (2001). Process Integration, Devices, and Structures and Emerging Research Devices. www.itrs.net. Accessed on 16 June 2013.

Manan, Z.A., Tan, Y.L. and Foo, D.C.Y. (2004). Targeting the minimum water flowrate using water cascade analysis technique, *AIChE Journal.* 50(12), 3169–3183.

Perry, R.H. and Green, D.W. (1998). *Perry's Chemical Engineers' Handbook Seventh Edition* New York, USA: McGraw-Hill.

Smith, R. (1995). *Chemical Process Design*, Chicester, UK: McGraw-Hill, Inc.

Wan Alwi, S.R. and Manan, Z.A. (2006). SHARPS – A new cost-ccreening technique to attain cost-effective minimum water utilisation network, *AIChE Journal*, 11 (52), 3981–3988.

Wan Alwi, S.R. and Manan, Z.A. (2008). A new holistic framework for cost effective minimum water network in industrial and urban sector, *Journal of Environmental Management*, 88, 219–252.

Wang, Y.P. and Smith, R. (1994). Wastewater Minimisation, *Chem. Eng. Sci.* 49, 981–1006.

Wang, Y.P. and Smith, R. (1995). Wastewater minimization with flowrate constraints, *Transactions of IChemE* Part A, 73, 889–904.

Water MATRIX© (2006). Software for Water Integration in Industries. Copyright of Universiti Teknologi Malaysia, Johor, Malaysia, www.cheme.utm.my/prospect, Accessed on 14 July 2013.

Wu, M., Sun, D. and Tay, J.H. (2004). Process-to-process recycling of high-purity water for semiconductor wafer backgrinding wastes, *Resources, Conservation and Recycling*, 41, 119–132.

*Note: For more information on the latest WPA developments, please refer to Chapter 9 'Sources of Further Information' at the end of this book.

9 Conclusions and sources of further information

This chapter has been divided into thematic parts dealing with Heat Integration, Total Site on the heat side, Mass and Water Integration on the mass and water side followed by some generic developments.

Heat Integration developed earlier and provided the base for the methodology that has been step-by-step extended in other directions such as Mass and Water Integration; however there have been some others such as Hydrogen Network Design and Management, Oxygen Pinch, Targeting for carbon emissions and footprints and also Property Pinch. The chapter concludes with an overview of the most prominent modelling tools suitable for applying Process Integration.

9.1 HEN targeting and synthesis

A critical review with an annotated bibliography for the literature on HENS up to year 2000 was published by Furman and Sahinidis (2002), while Anantharaman (2011) presented a HENS bibliography for the following period up to year 2010.

The HEN targeting and synthesis methodologies were developed and pioneered by the Department of Process Integration, UMIST (now the Centre for Process Integration, CEAS, The University of Manchester) in the late 1980s and 1990s. The main books published are as follows: Firstly the "red book" (Linnhoff et al., 1982, 1994), followed by a "blue book" Smith et al., 2000) updated later in an extended form (Smith, 2005). Another edition of Linnhoff's user's guide was published by Kemp (2007). Applications of Heat Integration in the food industry are presented by Klemeš and Perry (2007) included in an edited book of Klemeš et al. (2008). Some specialised parts providing more detailed instructions later in the retrofit methodology of Network Pinch can be found in (Asante and Zhu, 1997) and (Kemp, 2007).

The use of heat exchanger enhancements provides a cost-effective alternative to the purchase of new heat exchangers and/or investment in piping in retrofit design of Heat Recovery systems, as discussed by Wang et al. (2012). Key elements are low investment cost and avoiding fundamental structural modifications to the network. Bulatov (2005) has presented a retrofit optimisation framework for compact heat exchangers. Varbanov and Klemeš (2000) used the concept of heat load paths for the debottlenecking of HENS. A review of the recent developments in retrofit methodologies for HENS is provided by Smith et al. (2010).

9.2 Total Site Integration

Although Total Site Analysis and targeting has been used extensively for a number of years, there remains considerable academic and industrial interest in the area. Consequently, new journal and conference papers continue to be published in a number of research journals, although books are considerably less numerous.

Total Site methodology was firstly introduced by Dhole and Linnhoff (1993). Their work has been further developed by Raissi (1994) and extended by Klemeš et al. (1997).

Smith (2005) devoted in Chapter 23 in his book considerable space to the development of Total Sites and Total Site targeting including Total Site Profiles, Total Site Composites, and Site Utility Grand Composite Curves. He particularly pays attention to the use of the Site Utility Grand Composite Curves in relation to the cogeneration shaft work and the methods used to calculate the cogeneration shaft work.

Kemp (2007) also refers to Total Site Analysis in his chapter on Utilities, Heat and Power Systems. He briefly examines the construction of the Site Source and Site Sink Profiles.

A number of authors have extended the Total Site targeting methodology. Perry et al. (2008), made use of Total Site targeting using processes with different ΔT_{min} values to represent heat sources and sinks for the integration of small-scale chemical processes with a hospital complex and domestic buildings and office complexes. The work also examined the possibility of integrating renewable energy sources to reduce carbon-based emissions – for more details see Chapter 2.4. All previous Total Site research up to 2010 was overviewed in the recent book of Klemeš et al. (2010).

Matsuda et al. (2009) applied Total Site Analysis to the Kashima industrial area in Japan, which consists of 31 sites. They extracted data from the heaters and coolers from the Total Site in order to produce the Site Source and Site Sink Profiles.

Hackl et al. (2011) also applied Total Site Analysis to a cluster of five chemical companies producing a variety of products with variable supplies and demands for heating and cooling. They used the Total Site Profiles and Total Site Composite Curves to identify potential Heat Recovery and the identification for the need of a site-wide hot water circuit.

On the subject of cogeneration modelling, the recommended sources include the articles by Mavromatis and Kokossis (1998a,b) who proposed the steam turbine-based model for steam network optimisation and cogeneration targeting; Shang and Kokossis (2004) – who have developed on that basis a comprehensive model for steam level optimisation, also building upon Total Site (Klemeš et al., 1997) for steam levels integration; Varbanov et al. (2004a) on the further comprehensive model development and validation, also Varbanov et al. (2004b) for the improved targeting using the power-to-heat ratio and employing utility system optimisation to target processes for potentially beneficial process-level HEN retrofit.

9.3 Total Site methodology addressing variable energy supply and demand

In order to allow reduction of the carbon footprint from energy users, Perry et al. (2008) conceptually extended the TSHI by integrating greater community servicing within residential areas, with service and business centres as additional heat sinks and renewable energy as heat sources. The inherent variability in the heat supply and demand increases the difficulties in handling and controlling the system. Varbanov and Klemeš (2011) introduced the Time Slices into the Total Site description, with a heat storage system for accommodating the variations. In that paper, the Total Site Heat Cascade was introduced for visualising the heat flows across processes, the steam system, and the heat storage system.

Liew et al. (2012) introduced a numerical solution for the TSHI system to address the variable availabilities. This work presented a Total Site Heat Storage Cascade (TS-HSC) for addressing the heat storage facilities required by the TS system.

Nemet et al. (2012) discussed the approaches needed to maximise the usages of renewable energy sources with a fluctuating supply. They introduced a framework for integrating the production processes with solar energy, allowing a user to determine the amount of potential solar thermal energy that could be used within a process.

Varbanov et al. (2012) revisited the global minimum temperature difference (ΔT_{min}) used in the previous method. This work suggested that the values for ΔT_{min} should be specified individually for each site process. They demonstrated that the assumption of a global ΔT_{min} for the entire Total Site can be oversimplified, leading to potentially inadequate Total Site Heat Recovery targets. The modified targeting procedure allows obtaining more realistic Total Site Heat Recovery targets.

Liew et al. (2012) recently introduced a numerical method known as the Total Site Problem Table Algorithm (TS-PTA) for targeting the Total Site Utility Requirement, intended to allow automation of obtaining the TS targets. This method is easier to apply, and provides faster and more accurate results compared to the graphical approach, which has a tendency to include a graphical error during curve shifting. The Total Site Utility Distribution (TSUD) Table was also introduced to visualise the heat flows between the processes and the utility system. A numerical tool, the Total Site Sensitivity Table (TSST), for exploring site sensitivity was also proposed in the same paper. There are several mathematical models for the plant utility system planning process that incorporate the TSHI theory.

9.4 Utility system optimisation accounting for cogeneration

Mavromatis and Kokossis (1998a) introduced models for a steam system in the TSHI with two main objectives, the selection of the pressure steam levels and the determination of the operating unit configuration for the steam levels. Top-level analysis

(Varbanov et al., 2004b) is another mathematical modelling methodology that can be used for these concepts. This method allows for "scoping", i.e., selecting the site processes for targeting HI improvements. The current steam and power demands can be optimised and the potential benefit of reducing the steam demand can be assessed. A set of curves for the steam marginal prices can be produced for the system under consideration via top-level analysis.

Chen et al. (2011) proposed a systematic optimisation approach for designing a Steam Distribution Network (SDN) of steam systems in order to obtain an improved energy utilisation within the network. In this model, the operating conditions of the SDN were treated as design variables to be optimised. In another development on the TSHI, Bandyopadhyay et al. (2010) proposed a simplified methodology for targeting cogeneration potential based on the Salisbury (1942) approximation. This method is simple and linear, using the rigorous energy balance at the steam header.

Kapil et al. (2012) introduced a new model based on an isentropic expansion. These model results were favourable compared with the results from the detailed isentropic design methods and also included an optimisation study, which systematically determined the levels of the steam mains.

The TSST was proposed by Liew et al. (2012). This tool could be used to systematically determine the minimum and maximum boiler and cooling utilities' capacities under different operating conditions. However, this tool makes a major assumption, which treats the hot and cold utilities as the same types of utilities. This is because the steam generated cannot be used for satisfying the cooling requirement at the same steam level. Furthermore, the tool can be further developed for exploring the potential of cascading the excess energy at a high temperature in order to satisfy the LPS requirement during the process operational change scenario.

However, when implementing HI and TS HI in the industrial environment some important issues should be considered to deal with the model which is close to reality (Klemeš and Varbanov, 2010) and more recently Chew et al. (2013).

9.5 Maximum water recovery targeting and design

Water minimisation is a special case of Mass Integration that involves both mass transfer and non-mass transfer operations. There are two popular categories of approaches for water minimisation. These are the insight-based technique (also known as Water Pinch Analysis) and the mathematical optimisation technique. Professor Robin Smith and his co-workers at The University of Manchester, previously known as University of Manchester Institute of Science and Technology (UMIST), introduced the mass transfer-based WPA approach in the mid-1990s with the first work published in Wang and Smith (1994).

9.5.1 Recommended books for further reading

In his book, Pollution Prevention through Process Integration: Systematic Design Tools, El-Halwagi (1997) introduces the Pinch Analysis concept for Mass Integration that also covers the water minimisation case. The author proposes the use of a targeting technique involving water elimination, segregation, recycling, interception and sink/source manipulation.

Another good source of information on WPA with industrial applications is the book by Mann and Liu (1999). It covers water network planning, single and multiple contaminant targeting and network design for wastewater minimisation through water reuse. The book also examines the design of distributed effluent-treatment systems and reviews methods to target water regeneration, reuse, recycling and process changes. Detailed applications of the methods are demonstrated via a case study on a petrochemical complex.

The book by Smith (2005) covers water network targeting and design; water regeneration and techniques to handle process changes based on the WPA concept. The methods are however limited to systems involving single contaminants.

Klemeš et al. (2008) provide a comprehensive review on the legislation and techniques to improve the efficiency of water and energy use as well as wastewater treatment in the food industry. The book is divided into six parts. In Part 1, the book reviews the key drivers to improve water and energy management in food processing. In Part 2, methods to assess water and energy consumption and the strategies to efficiently design the water and energy systems arc presented. Part 3 covers good housekeeping procedures, measurement and process control to minimise water and energy consumption. Part 4 reviews the methods to minimise energy consumption in food processing, retail and waste treatment. Part 5 describes the water reuse and wastewater treatment in the food industry. In the last part, water and energy minimisation in several industrial sectors are presented.

Further information on the optimisation of water and energy use in food processing can be found in the book 'Waste management and co-product recovery in food processing' published by Waldron in two separate volumes. In Volume 1, Waldron (2007) reviews the latest developments in the area of waste management and co-product recovery in food processing in terms of legislative issues and technology for co-product separation and recovery. In Volume 2, Waldron (2009) discusses the life cycle analysis and closed-loop production system to minimise environmental impacts in food production. It also reviews methods to exploit co-products such as food and non-food ingredients.

Another overview on Water Integration is provided by Klemeš et al. (2010). The authors explain the use of the material recovery Pinch diagram with an application to a fruit juice case study, and the use of mathematical optimisation with application to a brewery case study.

In Foo et al. (2012), a compilation of several works on water minimisation techniques covering WPA, mathematical models, batch wastewater minimisation and adaptive swarm-based simulated annealing is presented in Section 2 of the book.

Foo (2012) has recently published a textbook on the latest Process Integration techniques for water minimisation, gas recovery and property integration. The book is divided into two parts. The first part of the book covers the insight-based, Pinch analysis graphical and algebraic targeting technique for direct reuse/recycle, regeneration, process changes, waste treatment, pre-treatment, batch and inter-plant. This is followed by a description of procedures for network design and evolution. The second part of the book is focused on the mathematical optimisation technique.

9.5.2 State of-the-art review

The first state-of-the-art critical review on water allocation and water treatment solutions that are based on the WPA and mathematical programming was published by Bagajewicz (2000). The roadmap towards zero liquid discharge and energy integrated solutions is also presented.

Foo (2009) presented an updated literature review only for the WPA approach for up to the year 2009, covering the targeting and network design stage from two angles, the *fixed load* and *fixed flowrate* problem. It also provides a review on the advanced targeting techniques for multiple freshwater sources, the threshold problem, impure freshwater source(s), water regeneration, wastewater treatment and total water network. The author also provided some commentaries on the future research directions that may involve other areas of resource conservation problems including the multiple impurity problems, simultaneous targeting and design of water networks, simultaneous heat and water recovery, water network retrofit and interplant Water Integration.

A very detailed review of water network design methods with literature annotations from journals and conferences up to mid-2009 was presented by Jeżowski (2010). The description of each paper is divided into scope, method and special features. The paper presents reviews on the total water network (TWN), water-using network (WUN), and wastewater treatment network (WWTN) problems for continuous operations, batchwise operations, Simultaneous Heat and Water Integration, and uncertain data and interplant integrations. A total of 264 literature annotations until 2009 were presented by the author.

A review on the WPA, mathematical programming techniques and combined water-energy minimisation for up to the year 2012 was done by Klemeš (2012). A review on the water footprint and life cycle analysis, as well as a short overview of several case studies, were also included.

The WPA technique has seen further developments in 2013. Parand et al. (2013) recently proposed an improved methodology for Water Cascade Analysis and Material Recovery Pinch Diagram for the threshold problem with external utilities. Shenoy and Shenoy (2013) developed the Unified Targeting Algorithm (UTA) and the Nearest-Neighbours Algorithm (NNA) to target and design cooling water networks.

Liu et al. (2013) presented a new insight-based method to design the distributed effluent treatment systems for systems with multiple contaminants. The new method can reduce unreasoning stream-mixing in the design of distributed wastewater treatment networks. Wastewater degradation caused by unreasoning stream-mixing will increase the total treatment flowrate, which will subsequently increase the treatment cost.

Agana et al. (2013) demonstrated the application of an in-series integrated water management strategy for two large Australian manufacturing companies that produce different products and generate different wastewater types. The integrated strategy to identify water conservation opportunities consists of water audit, Pinch Analysis and a membrane process application. This strategy can be used as a guide for other manufacturing sites to develop their water management plans.

Due to the growing concern about water scarcity as well as water pollution, there has been renewed interest shown by researchers in solving water minimisation problems. Consequently, more works on water minimisation are expected to be developed and published in the near future by researchers across the globe.

9.6 Analysing the designs of isolated energy systems

An isolated energy system serves the energy demand of a location by providing energy near its point of utilisation and can also provide guidance for highly self-sufficient regions. Isolated energy systems, independent of the national grid, are employed for remote electrification. A very good description has been provided by Bandyopadhyay (2013): Besides electrical systems, similar isolated energy systems can also meet the thermal demand. The advantages of such systems are their low fuel cost, non-polluting capacities, modularity, and ease of extensibility. The disadvantages are associated with them mostly being unpredictable and the low energy densities (kW/m^2) of different renewable resources and the higher capital investments. It is also important to provide an energy storage system for improving the rate of utilisation of the renewable energy sources. The performance depends upon proper sizing of the overall system. Optimisation of the entire system is needed for sizing which satisfies certain cost and reliability criteria. Pinch Analysis is applied to the design and optimisation of isolated renewable energy systems.

The cumulative energy generated and required may be plotted on a time vs. energy axis. These curves are equivalent to the Composite Curves of Pinch Analysis. Typical Composite Curves, Supply Composite Curves, and Demand Composite Curves, were plotted by Bandyopadhyay (2011). The energy is equivalent to the heat transfer and the time is equivalent to the temperatures of the Composite Curves of Pinch Analysis and even more to Total Site problems. Energy generated in the renewable generators can be supplied to the energy required by the load. Energy generated during or before the demand can be supplied; energy generated after the demand cannot be supplied to the demand. The Energy Supply Composite Curve has to lie above the Energy Demand Composite Curve. Fulfilling this condition, the energy Supply Composite Curve may be shifted upwards, until it touches and lies completely above the Demand Composite Curve. Another work in a similar direction has been presented by Ho et al. (2013).

9.7 PI contribution to supply chain development

The introduction of supply chain design or supply chain management aims at minimising the total supply chain cost for meeting a present and known demand (Shapiro, 2001).

However, the conventional approach to supply chain modelling is outdated, mainly due to the lack of consideration of carbon emissions. According to EIA (2005) the transportation and industrial sectors appear to be the sectors with the largest CO_2 emissions. The techniques derived from Pinch Analysis, which provides a clear insight into the PI concept, bring also a significant contribution to the development of a supply chain design (Lam et al., 2011). The supply chain design using PI provides an analysis pathway. The combination of techniques allows for a supply chain design with an offset between cost and environmental impact for a compromise volume of carbon emission.

Lam et al. (2013) provided an overview of conventional supply chain applications follow by the advantages that PI has brought to the supply chain regarding the aspects of social, economic, environmental impact, and resources utilisation.

A supply chain system engages production facilities, warehouses, distributors, suppliers, retailers and customers. One of them usually acts as a 'bottleneck' or Pinch that imposes limitations on the system's capacity. The PI of the supply network provides an alternative graphical solution of 'quality' versus 'quantity'. The graphical plot through Pinch methodology allows for identification of the Pinch location. The 'debottlenecking' process removes the system's restriction, which may lead to an improvement in the supply chain system's performance.

9.8 Hydrogen networks design and management

The further evolution of Pinch Technology extended Mass Integration to hydrogen management systems. Environmental considerations and resource availability constraints, such as low aromatics gasoline, low sulphur diesel, shift to heavier crude oil, more upgrading, lead to increased demand for hydrogen in refineries and at the same time reduce the sources of hydrogen traditionally available to refineries. This can result in a deficit in the hydrogen balance of a refinery, and in the need for investment in additional hydrogen production facilities or importing it from outside suppliers. If the available hydrogen resources can be used more efficiently, the requirements of additional production capacity or imports can be decreased.

In one of the early works, Alves (1999) proposed a Pinch Approach to targeting the minimum hydrogen utility. This technology was based on an analogy with process Heat Recovery. Similar to distribution of energy resources in a plant which can be analysed and designed by using Pinch Technology, distribution of hydrogen resources in a refinery has a number of potential sources, each capable of producing a different amount of hydrogen, and a number of hydrogen sinks with different requirements. However, the engineer has more flexibility in deciding the hydrogen loads of individual units by varying throughputs of units and operating many processes over a range of conditions. This leads to considerable potential for optimisation of refinery performance. Using the Hydrogen Pinch Analysis the engineer can make best use of hydrogen resources in order to meet new demands and improve profitability.

A very good example is the recent work by Liu et al. (2013), which examines the targeting of hydrogen networks with purification. A more detailed description of Hydrogen Pinch can be found elsewhere (Klemeš et al., 2010).

9.9 Oxygen Pinch Analysis

Another extension of the Pinch concept was the Oxygen Pinch Analysis (Zhelev and Ntlhakana, 1999). Their idea of Oxygen Pinch Analysis is to analyse the system to come up with targets prior to designing a system with minimum oxygen consumption by the micro-organisms for waste degradation. The next step is to design a flowsheet and operating conditions providing the target. In most of the cases the oxygen is supplied to the micro-organisms through agitation. Aeration requires energy, so eventually the analysis based on Oxygen Pinch principles leads to the original application associated with energy conservation.

The Oxygen Pinch concept focuses on Mass Transfer Pinch analysis but exceeds the classical Pinch targets expectations. Using the chemical oxygen demand as a common ground for a range organic contaminants it sets quantitative targets (oxygen solubility, residence time, oxidation energy load), as well as additional qualitative targets, namely the growth rate directly addressing the micro-organisms' age and health (Zhelev and Bhaw, 2000).

The main idea of the method is the oxidation/aeration energy minimisation required for biodegradation of organic waste through aerobic digestion. The targeting procedure is based on a composite plot of the chemical oxygen demand (COD) against the dilution rate, which builds on the Monod model of biodegradation (Monod, 1949).

The reason why the researchers (Zhelev and Bhaw, 2000) applied COD (organic load or chemical oxygen demand) vs. the dilution rate (D) is to reverse the graph and build the limiting concentration profile from process streams, not from the dissolved oxygen stream, which might cause problems in data collection and analysis. The analysis of information is followed by matching the oxygen supply line to the Composite Curve (touching it at the point of Pinch) and provides the following pre-design information:

1. Micro-organisms' growth rate;
2. Oxygen solubility;
3. Residence time;
4. Oxidation energy load.

The method brings us back to the energy saving domain in Pinch Technology development, which later was developed in the environmental area – Water Pinch. In the Oxygen Pinch approach, it gets back to energy but is accompanied by extra information concerning environmental issues. An important contribution of this method is its ability to target in parallel with the concentration and the total energy required, a quality characteristic – the micro-organisms' health assessed through their reproduction rate. The design guidelines follow the Water Pinch analogy. This method provides an improved tool for design of cost-efficient biotreatment processes.

9.10 Pressure drop considerations and heat transfer enhancement in Process Integration

Various factors, such as flowrate, composition, temperature and phase can affect heat capacity C_p. Pressure is another very important factor to be taken into account. Back in 1990, Polley et al. (1990) extended the targeting procedure by considering pressure drops. They used the relationship between the pressure drop ΔP, heat transfer coefficient h, and the heat transfer area A:

$$\Delta P = KAh^m. \tag{9.1}$$

Instead of specifying the heat transfer coefficients for streams, the allowable pressure drop was specified for each stream. The heat transfer coefficients for the streams are calculated iteratively to minimise the total area. The target area procedure is modified based on the fixed pressure drops rather than fixed film coefficients.

Ciric and Floudas (1989) suggested a mathematical programming-based two-stage approach. It included a match selection stage and an optimisation stage. The match selection stage used an MILP transhipment model to select process stream matches and match assignments. The optimisation stage used NLP formulation to optimise the match order and flow configuration of matches.

Nie and Zhu (1999) developed a strategy for considering the pressure drop for HEN retrofit. They assume that additional area should be concentrated on a small number of units to minimise the piping and civil work. The optimisation procedure consists of two stages. First stage: Screening for a small number of units which require additional area. Second stage: Considering a serial or parallel shell arrangements for those units.

The topology change options are first determined by applying the Network Pinch Method (Asante and Zhu, 1997). Then their two-stage optimisation is used for area distribution and shell arrangement with pressure drop constraints. Area distribution and shell arrangement are two most common options which have a main effect on pressure drop.

Václavek et al. (2003) classified in more detail when pressure plays an important role in Energy Process Integration. Rather than individual process streams available for Heat Recovery, whole combinations of process streams (tracks) are considered. They formulated some heuristic rules.

Aspelund et al. (2007) described a new methodology to account for pressure drops for process synthesis which extends traditional Pinch Analysis with exergy calculations: Extended Pinch Analysis and Design (ExPAnD). The authors focus on the thermo-mechanical exergy, which is the sum of pressure and temperature-based exergy.

Compared with traditional Pinch Analysis, the problem which Aspelund et al. (2007) consider (subambient process-based) is much more complex with a large number of alternatives for manipulation and integration of streams. Aspelund et al. (2007) also provide a number of heuristics (general and for specific streams) that complement the ExPAnD methodology. In the further development, Aspelund and Gundersen (2007), using the concept of Attainable Region, suggest a graphical representation of all possible Composite Curves for a pressurised cold stream below ambient to include the cooling effect when the stream is expanded to its target pressure. The Attainable Region is an addition to the ExPAnD methodology, a new tool for Process Synthesis extending Pinch Analysis by explicitly accounting for pressure and including exergy calculations. The methodology shows great potential for minimising total shaft work in subambient processes.

To account for debottlenecking retrofit (which should be distinguished from the energy saving retrofit) Panjeshahi and Tahouni (2008) suggested a pressure drop optimisation method. Its main stages are as follows.

1. Simulation of the existing process operating at the desired increased throughput. Additional utility is used to restore the required temperatures in the process.
2. Area efficiency specification in the existing network after increasing throughput using area-energy plot. A new virtual area is introduced, which is named the pseudonetwork.

Zhu et al. (2000) suggested a heat transfer enhancement procedure for HEN retrofit. The methodology has two stages – the targeting stage and selection. The approach developed provides a quick way to determine the application of enhancement in the conceptual design.

The method is based on the results of Network Pinch Analysis. The limitation of the method is that heat transfer enhancement is only used to take the place of additional area.

Smith et al. (2013) provide an introduction to the enhancement methods of commercial shell-and-tube heat exchangers and their engineering application. The key aspects of enhancing shell-and-tube heat exchangers in Heat Exchanger Network retrofit are presented and illustrated with examples involving both energy saving and fouling considerations. This analysis has been based on a series of works by Pan. Pan at al. (2011a) studied the improvement of energy recovery in Heat Exchanger Networks with intensified tube-side heat transfer. Pan et al. (2011b) introduced an optimisation method for retrofitting Heat Exchanger Networks with intensified heat transfer. Pan et al. (2012) presented MILP-based (Mixed Integer Linear Programming) iterative method for the retrofit of Heat Exchanger Networks with intensified heat transfer and most recently Pan et al. (2013) developed an optimisation procedure for the retrofit of large-scale Heat Exchanger Networks with intensified heat transfer.

Heat transfer can be intensified in shell-and-tube heat exchangers for improved heat transfer performance. Such intensification has been widely studied from the point of view of individual heat exchangers. The use of heat transfer enhancement in Process Integration has many benefits. First, enhanced heat exchangers require less heat transfer area for a given heat duty because of higher heat transfer coefficients. Second, the heat transfer capacity for the given heat exchanger can be increased without changing physical size of the exchanger. Third, the use of enhancement can reduce pumping requirement in some cases as enhanced heat exchangers can achieve higher overall heat transfer coefficients with lower velocities, which may lead to lower friction losses. Using enhancement techniques has practical advantages in Heat Exchanger Network retrofit, as it can avoid modification of the network structure. The implementation of enhancement is a relatively simple task that can be easily achieved within a normal maintenance period, allowing production losses to be kept to a minimum level and no civil engineering required.

9.11 Computational and modelling tools suitable for applying PI

Process Integration, modelling, and intensification problems are complex in terms of scale and relationships. Solving is very much aided by relevant computer software. There are a large number of efficient tools available, each with its particular advantages. Here the main tools used in the research and applied calculations are discussed. In (Klemeš et al., 2010) – "Sustainability in the Process Industry – Integration and Optimisation" a comprehensive list of software tools is provided. However, from these only a small number are directly suitable for applying PI.

9.11.1 Heat and power PI applications

There are dedicated tools for Heat Integration. The earlier (legacy) tool is named SPRINT. This is a software package for designing Heat Recovery systems for individual processes (SPRINT, 2012). This software provides energy targets and optimises the choice of utilities for an individual process; it also performs Heat Exchanger Network (HEN) design automatically for the selected utilities. Both new design and retrofit are carried out automatically, though the designer maintains control over network complexity. SPRINT can also be used for HEN operational optimisation. The recent successor of SPRINT is HEAT-int (2013). This is a product of Process Integration Ltd. The program is used to improve the energy performance of individual processes on a site. It is the next generation development of the SPRINT software (by the same team of developers) to a commercial standard, and offers more user-friendly interface features.

Another very reputable software package for Process Integration modelling is SuperTarget (KBC, 2013). SuperTarget is mainly used to improve Heat Integration in new design and retrofit projects by reducing operating cost and optimally targeting capital investment. It is also a tool for day-to-day application by novices or occasional users, and it makes Pinch Analysis a routine part of process design.

STAR is a software package for the optimisation of site utility and cogeneration systems (STAR, 2012). It analyses the interactions between site processes and the utility system, local fired heaters, and cooling systems. The analysis can be used to reduce energy costs and to plan infrastructure adjustments. STAR can also be used to investigate flue gas emissions, which must often be reduced to meet tighter environmental regulations. The STAR package incorporates several tools, as described next. One of the advanced procedures possible with STAR is top-level analysis, which is used for identifying potential processes for Heat Integration retrofit based on limited information from the utility system only, saving investment and time to industrial companies.

Like HEAT-int, SITE-int (2013) is a product of Process Integration Ltd. It is the successor of STAR. SITE-int is a state-of-the-art software package for the design, optimisation, and integration of site utility systems in process industries. Its main features are similar to those of STAR, but updated and commercial support is included.

One particularly interesting tool is Aspen Energy Analyzer (Aspentech, 2013). It is intended to provide an environment for optimal Heat Exchanger Network design and retrofit. In this sense it is a direct competitor to HEAT-int and SPRINT. It also provides integration and interaction with other Aspentech tools such as Aspen Plus and Aspen HYSYS.

It is also possible to perform the targeting parts and other calculations using spreadsheets. An example is the spreadsheet tool developed as an addendum to the PI book by Kemp (2007).

9.11.2 Water Pinch software

There is much commercial software available that can optimise water networks in industry either based on the WPA or mathematical modelling. The software can help users with limited knowledge of the technology to perform water minimisation in their facilities. There follows are a brief review of available software.

WaterTarget (2013) by KBC Energy Services consists of two standalone programs, i.e. WaterTracker and Water Pinch. Water Tracker can be used to develop the site water balances and reduce time in data collection. It helps the user to identify where to measure next, how to resolve data conflicts, and can reconcile data points. Water Pinch allows data to be imported from WaterTracker. Flowsheets can be developed and constraints for water reuse can be specified. Distances, pump and treatment unit capacity can also be included in the software. The software also provides operating cost (water purchase, treatment, pumping and discharge) analysis, optimal network connection analysis and sensitivity analysis.

ASPEN Water (2013) has been developed by Aspen Technology Inc., USA. A user can develop a water flowsheet with the respective unit operations by using this software. Software can be linked to ASPEN Properties to model the water chemistry of the plant. It can also perform data reconciliation to provide estimates of flow and compositions. This software uses optimisation techniques to minimise water usage and minimise water network costs. The software allows users to determine the water costs, connection costs, fixed and variable regeneration costs and discharge costs in its model. ASPEN Water can also perform contaminant sensitivity analysis, process modifications, regeneration, wastewater treatment selection and optimisation.

Water (2013) has been developed by the Centre for Process Integration, The University of Manchester. It can minimise freshwater consumption and wastewater generation through the maximum reuse and regeneration. The software can cater for systems with multiple contaminants and also multiple freshwater sources. It can automatically design the water reuse and effluent treatment networks. The software also provides a trade-off analysis between freshwater, effluent treatment and pipework/sewer cost.

Optimal Water (2013) (previously known as Water MATRIX) is the software developed by Process Systems Engineering Centre (PROSPECT), Universiti Teknologi Malaysia. The software can perform minimum water targeting for the single contaminant system. It uses the Water Cascade Analysis method developed by Manan et al. (2004) and the Balanced Composite Curves as well as Water Surplus Diagram method by Hallale (2002). This software can automatically generate water network designs based on the Source and Sink Mapping Diagram (Polley and Polley, 1998). The software is also capable of performing simultaneous water and energy minimisation and minimum regeneration targeting.

OptWatNet (Teles et al., 2007) has been developed by DMS/Instituto Nacional de Engenharia Tecnologia e Inovação, Portugal. It can design water-using networks with multiple water sources and contaminants. The software uses LP/NLP models and solves the water minimisation problem by using GAMS. The user interface is developed by using Microsoft Visual Studio 2005.

Water Design (2013) has been developed by Virginia Polytechnic Institute and State University. It employs the Water Composite Curves method to target the maximum water recovery, and can generate the optimum water network design. The software tool adopts the method developed by Wang and Smith (1994) that is limited to single contaminant systems.

9.12 Challenges and recent developments in Pinch-based PI

Bandyopadhyay (2006) presented another extension to the Source Composite Curve for waste reduction and Atkins et al. (2010) presented a Carbon Emissions Pinch Analysis (CEPA) for emissions reduction in the electricity sector. A graphical targeting tool for the planning of carbon capture and storage issues was analysed by Ooi et al. (2012).

Gerber et al. (2013) defined optimal configurations of geothermal systems using process design and PI techniques. The problem of soft data in PI was studied and industrially applied by Walmsley et al. (2013); an important issue, which deserves extended attention in PI problems as well as data reconciliation (Manenti et al, 2011).

PI issues not covered in this overview, but still important, are targeting and design of batch processes, see e.g. Foo et al. (2012) and cost-effective Heat Exchanger Network design for non-continuous processes by Morrison et al. (2012).

The most recent overviews of PI, in most cases related to sustainability, include Klemeš et al. (2010): Sustainability in the Process Industry – Integration and Optimisation, El-Halwagi (2012): Sustainable Design through PI – Fundamentals and Applications to Industrial Pollution Prevention, Resource Conservation, and Profitability Enhancement, Foo (2012): PI for Resource Conservation and Foo (2013): A Generalised Guideline for Process Changes for Resource Conservation Networks. Most recently an edited Handbook on PI was published to which most leading researchers contributed (Klemeš, ed. 2013).

This list can never be fully comprehensive and very substantial research results are being published virtually every week. The research and new results in all directions of Process Integration continues and its growth has even been accelerating. This provides very strong evidence that Process Integration methodology, both based on Pinch and complemented by mathematical programming, has still not reached its saturation point and many extensions are still to be envisaged.

To closely follow the most recent developments, the PRES Conferences (2013) are probably the best option. The 16th Conference on Process Integration, Modelling and Optimisation for Energy Saving and Pollution Reduction – PRES'13 took place in Rhodes, Greece; the 17th Conference PRES 2014 in Prague, the Czech Republic and the

18th PRES'14 in Malaysia, Sarawak, Borneo. Valuable contributions have been regularly selected from PRES conferences and published in Chemical Engineering Transactions (2013) and also as special issues of the Journal of Cleaner Production, Energy, Applied Thermal Engineering, Cleaner Technologies and Environmental Policy, Theoretical Foundation of Chemical Engineering and Resources, Conservation and Recycling.

References

Agana, B.A., Reeve, D. and Orbell, J.D. (2013). An approach to industrial water conservation – A case study involving two large manufacturing companies based in Australia, *Journal of Environmental Management*, 114, 445–460.

Anantharaman, R. (2011). *Energy efficiency in process plants with emphasis on heat exchanger networks – Optimization, thermodynamics and insight*, PhD Thesis, Norwegian University of Science and Technology, Trondheim, Norway.

Asante, N.D.K. and Zhu, X.X. (1997). An automated and interactive approach for heat exchanger network retrofit, *Chemical Engineering Research and Design*, 75(part A), 349–360.

ASPEN Properties (2013). Aspen Technology, Inc. www.aspentech.com/products/aspen-properties. aspx, accessed on 7 July 2013.

AspenTech (2013). Aspen Energy Analyzer – AspenTech. www.aspentech.com/products/ aspen-hx-net.aspx, accessed 9 August 2013.

ASPEN Water (2013). Aspen Technology, Inc., www.aspentech.com/brochures/Aspen_Water_ Brochure.pdf, accessed on 7 July 2013.

Bagajewicz, M.J., Rivas, M. and Savelski, M.J. (2000). A robust method to obtain optimal and sub-optimal design and retrofit solutions of water utilization systems with multiple contaminants in process plants, *Computers and Chemical Engineering*, 24, 1461–1466.

Bandyopadhyay, S. (2013). Applications of Pinch Analysis in the design of isolated energy systems, in: Klemeš (ed.) *Process Integration Handbook*, Cambridge, UK: Woodhead Publishing, ISBN-13: 978 0 85709 593 0.

Bandyopadhyay, S. (2011). Design of renewable energy systems incorporating uncertainties through pinch analysis, *Computer Aided Chemical Engineering*, Elsevier, 29, 1994–1998.

Bandyopadhyay, S., Varghese, J. and Bansal, V. (2010). Targeting for cogeneration potential through total site integration, *Applied Thermal Engineering*, 30(1), 6–14.

Bulatov, I. (2005). Retrofit optimization framework for compact heat exchangers, *Heat Transfer Engineering*, 26, 4–14.

Chemical Engineering Transactions www.aidic.it/cet accessed 12 August 2013.

Chew, K.H., Klemeš, J.J., Wan Alwi, S.R. and Manan, Z.A. (2013). Industrial implementation issues of Total Site Heat Integration, *Applied Thermal Engineering*, doi: 10.1016/j.applthermaleng.2013.03.014

El-Halwagi, M.M. (1997). *Pollution Prevention through Process Integration: Systematic Design Tools*, San Diego, USA: Academic Press.

El-Halwagi, M.M. (2012), *Sustainable Design through Process Integration – Fundamentals and Applications to Industrial Pollution Prevention, Resource Conservation, and Profitability Enhancement*, Amsterdam, The Netherlands: Elsevier. www.knovel.com/web/ portal/browse/ display?_EXT_KNOVEL_DISPLAY_bookid=5170&VerticalID=0, accessed 06/07/2013.

Foo, D.C.Y. (2009). State-of-the-art review of pinch analysis techniques for water network synthesis, *Industrial and Engineering Chemistry Research*, 48 (11), 5125–5159.

Foo, D.C.Y. (2012). *Process Integration for Resource Conservation*, Boca Raton, Florida, USA: CRC Press.

Foo, D.C.Y. (2013), A generalised guideline for process changes for resource conservation networks, *Clean Technologies and Environmental Policy*, 15 (1), 45–53.

Foo, D.C.Y., El-Halwagi, M.M. and Tan, R.R. (2012). *Advances in Process Systems Engineering – Vol. 3. Recent Advances in Sustainable Process Design and Optimization*, Singapore: World Scientific.

Furman K.C. and Sahinidis N.V. (2002). A critical review and annotated bibliography for heat exchanger network synthesis in the 20th century, *Industrial & Engineering Chemistry Research*, 41, 2335–2370.

Hackl, R., Andersson, E. and Harvey, S. (2011). Targeting for energy efficiency and improved energy collaboration between different companies using total site analysis (TSA), *Energy*, 36, 4609–4615.

HEAT-int (2013). HEAT-int | Process Integration Limited. www.processint.com/chemical-industrial-software/heat-int, accessed 02/08/2013.

Ho, W.S., Hashim, H., Lim, J. S. and Klemeš, J.J. (2013). Combined design and load shifting for distributed energy system, *Clean Technologies and Environmental Policy*, 15 (3), 433–444.

Jeżowski, J. (2010). Review of water network design methods with literature annotations, *Industrial and Engineering Chemistry Research*, 49 (10), 4475–4516.

KBC (2013). SuperTarget™ – KBC Advanced Technologies plc. www.kbcat.com/energy-utilities-software/supertarget, accessed 02/08/2013.

Kemp, I.C. (2007). *Pinch Analysis and Process Integration. A User Guide on Process Integration for Efficient Use of Energy*, Amsterdam, The Netherlands: Elsevier, (authors of the first edition: Linnhoff, B., Townsend, D.W., Boland, D., Hewitt, G.F., Thomas, B.E.A., Guy, A.R. and Marsland, R.H. (1982, latest edition 1994), A User Guide on Process Integration for the Efficient Use of Energy, Rugby, UK: IChemE).

Klemeš, J. J. (2012). Industrial water recycle/reuse, *Current Opinion in Chemical Engineering*, 1, 238–245.

Klemeš, J., Dhole, V.R., Raissi, K., Perry, S.J. and Puigjaner, L. (1997). Targeting and design methodology for reduction of fuelp Power and CO_2 on Total Sites, *Applied Thermal Engineering*, 7, 993–1003.

Klemeš, J., Friedler, F., Bulatov, I. and Varbanov, P. (2010). Sustainability in the Process Industry: Integration and Optimization, New York, USA: McGraw Hill Companies Inc.

Klemeš, J. and Perry, S. (2007). Process optimisation to minimise energy use in food processing, in: Waldron K. (ed.), *Handbook of Waste Management and Co-product Recovery in Food Processing*, 1, pp. 59–89, Cambridge, UK: Woodhead.

Klemeš, J., Smith, R. and Kim, J.-K. (eds.) (2008). *Handbook of Water and Energy Management in Food Processing*. Cambridge, UK: Woodhead.

Lam, H.L., Klemeš, J.J., Kravanja, Z. and Varbanov, P. (2011). Software tools overview: Process Integration, modelling and optimisation for energy saving and pollution reduction, *Asia-Pacific Journal of Chemical Engineering*, 6(5), 696–712.

Lam, H.L., Klemeš, J.J., Varbanov, P.S. and Kravanja, Z. (2013) P-graph synthesis of open structure biomass networks, *Industrial and Engineering Chemistry Research*, 52 (1), 172–180.

Liew, P.Y., Wan Alwi, S. R., Varbanov, P.S., Manan, Z.A. and Klemeš, J.J. (2012). A numerical technique for Total Site sensitivity analysis, *Applied Thermal Engineering*, 40, 397–408.

Linnhoff, B., Townsend, D.W., Boland, D., Hewitt, G.F., Thomas, B.E.A., Guy, A.R. and Marsland, R.H. (1982). *A User Guide to Process Integration for the Efficient Use of Energy*, Rugby, UK: IChemE, latest edition 1994.

Linnhoff, B. and Vredeveld, D.R. (1984). Pinch technology has come of age, *Chemical Engineering Progress*, 80(7), 33–40.

Liu, Z., Shi, J. and Liu, Z. (2013). Design of distributed wastewater treatment systems with multiple contaminants, *Chemical Engineering Journal*, 228, 381–391.

Liu, G., Li, H., Feng, X., Deng, C. and Chu, K.H. (2013). A conceptual method for targeting the maximum purification feed flow rate of hydrogen network, *Chemical Engineering Science*, 88, 33–47.

Manenti, F., Grottoli, M.G., Pierucci, S. (2011). Online data reconciliation with poor-redundancy systems, *Industrial & Engineering Chemistry Research*, 50(24), 14105–14114.

Mann, J.G., and Liu, Y.A. (1999). *Industrial Water Reuse and Wastewater Minimization*, New York, USA: McGraw Hill.

Matsuda, K., Hirochi, Y., Tatsumi, H. and Shire, T. (2009). Applying heat integration total site based pinch technology to a large industrial area in Japan to further improve performance of highly efficient process plants, *Energy*, 34(10), 1687–1692.

Mavromatis, S.P. and Kokossis, A.C. (1998a). Conceptual optimisation of utility networks for operational variations – I. Targets and level optimisation, *Chem. Eng. Sci.*, 53(8), 1585–1608.

Mavromatis, S.P. and Kokossis, A.C. (1998b). Conceptual optimisation of utility networks for operational variations – II. Network development and optimisation, *Chemical Engineering Science*, 53(8), 1609–1630.

Morrison, A.S., Atkins, M.J. and Walmsley, M.R.W. (2012). Ensuring cost-effective heat exchanger network design for non-continuous processes, *Chemical Engineering Transactions*, 29, 295–300.

Ooi, R.E.H., Foo, D.C.Y., Ng, D.K.S. and Tan, R.R. (2012). Graphical targeting tool for the planning of carbon capture and storage, *Chemical Engineering Transactions*, 29, 415–420.

Optimal Water (2013). Process Systems Engineering Centre (PROSPECT), Universiti Teknologi Malaysia. www.cheme.utm.my/prospect, accessed on 7 July 2013.

Pan, M., Bulatov, I., Smith, R. and Kim, J.-K. (2011a). Improving energy recovery in heat exchanger network with intensified tube-side heat transfer, *Chemical Engineering Transactions*, 25, 375–380, DOI: 10.3303/CET1125063.

Pan, M., Bulatov, I., Smith, R. and Kim, J.-K. (2011b). Novel optimization method for retrofitting heat exchanger networks with intensified heat transfer, *Computer Aided Chemical Engineering*, 29, 1864–1868, DOI: 10.1016/B978-0-444-54298-4.50151-3.

Pan, M., Bulatov, I., Smith, R. and Kim, J.-K. (2012). Novel MILP-based iterative method for the retrofit of heat exchanger networks with intensified heat transfer, *Computers and Chemical Engineering*, 42, 263–276, DOI: 10.1016/j.compchemeng.2012.02.002.

Pan, M., Bulatov, I., Smith, R. and Kim, J.-K. (2013). Optimisation for the retrofit of large scale heat exchanger networks with different intensified heat transfer techniques, *Applied Thermal Engineering*, 53(2) 373–386.

Parand, R., Yao, H.M, Tadé, M.O. and Pareek, V. (2013). Targeting water utilities for the threshold problem without waste discharge, *Chemical Engineering Research and Design*. doi:10.1016/j. cherd.2013.05.004.

Perry, S., Klemeš, J. and Bulatov, I. (2008). Integrating waste and renewable energy to reduce the carbon footprint of locally integrated energy sectors, *Energy*, 33 (10) 1489–1497.

PRES Conference www.confrencepres.com accessed 12 August 2013.

Shenoy, A.U. and Shenoy, U.V. (2013). Targeting and design of CWNs (cooling water networks), *Energy*, 55, 1033–1043.

SITE-int (2013). HEAT-int | Process Integration Limited. www.processint.com/chemical-industrial-software/site-int, accessed 02/08/2013.

Smith, R (2005). *Chemical Process Design and Integration*, Chichester, UK: John Wiley & Sons.

Smith, R., Jobson, M. and Chen, L. (2010). Recent developments in the retrofit of heat exchanger networks, *Applied Thermal Engineering*, 30, 2281–2289.

Smith, R., Pan, M. and Bulatov, I. (2013). Heat Exchanger Networks with Heat Transfer Enhancement, in: Klemeš J.J., (ed.) (2013). *Process Integration Handbook*, Cambridge, UK: Woodhead Publishing, pp. 966–1035, DOI: 10.1533/9780857097255.5.966.

Shang, Z. and Kokossis, A. (2004). A transhipment model for the optimisation of steam levels of total site utility system for multiperiod operation, *Computers and Chemical Engineering*, 28(9), 1673–1688.

STAR (2012). Process Integration Software (Centre for Process Integration, School of Chemical Engineering and Analytical Science, The University of Manchester, U.K.). www.ceas.manchester. ac.uk/research/centres/ centreforprocessintegration/software/packages/star, accessed 26/10/2009.

SPRINT (2012). Process Integration Software (Centre for Process Integration, School of Chemical Engineering and Analytical Science, The University of Manchester, U.K.). www.ceas.manchester. ac.uk/research/centres/ centreforprocessintegration/software/packages/sprint, accessed 26/10/2009.

Teles, J.P., Castro, P.M. and Novais, A.Q. (2007). OptWatNet – A software for the optimal design of water-using networks with multi-contaminants, *Computer Aided Chemical Engineering*, 24, 497–502.

Varbanov, P.S., Doyle, S. and Smith, R. (2004a). Modelling and optimisation of utility systems, *Trans IChemE, Chem Eng Res Des*, 82(A5), 561–578.

Varbanov, P., Perry, S., Makwana, Y., Zhu, X.X. and Smith, R. (2004b), Top-level analysis of site utility systems, *Transaction IChemE, Chemical Engineering Research Design*, 82(A6), 784–795.

Varbanov, P.S, Fodor, Z., Klemeš, J.J. (2012). Total Site targeting with process specific minimum temperature difference (ΔT_{min}), *Energy*, 44(1), 20–28, doi:10.1016/j.energy.2011.12.025.

Varbanov, P.S. and Klemeš, J. (2000). Rules for paths construction for HENs debottlenecking, *Applied Thermal Engineering*, 20, 1409–1420.

Varbanov, P. and Klemeš, J. (2011). Integration and management of renewables into Total Sites with variable supply and demand, *Computers and Chemical Engineering*, 35(9), 1815–1826.

Waldron, K. (ed.) (2007). *Waste Management and Co-product Recovery in Food Processing*, Cambridge, UK: Woodhead Publishing Limited, pp. 90–118.

Waldron, K. (ed.) (2009). *Waste Management and Co-product Recovery in Food Processing*. Vol 2, Cambridge, UK: Woodhead Publishing Limited, 391(6), pp. 134–168. ISBN: 978 1 84569.

Wang, Y., Pan, M., Bulatov, I., Smith, R. and Kim, J.-K. (2012). Application of intensified heat transfer for the retrofit of heat exchanger networks, *Applied Energy*, 89, 45–59.

Water Design (2013). Virginia Polytechnic Institute and State University Virginia. www.design.che. vt.edu/waterdesign/waterdesign.html, accessed 7/7/2013.

Water (2013). Centre for Process Integration, The University of Manchester. www.ceas.manchester. ac.uk/media/eps/schoolofchemicalengineeringandanalyticalscience/content/researchall/ centres/processintegration/WATER.pdf, accessed 7/7/2013.

WaterTarget (2013). KBC Energy Services. www.kbcat.com/Products-and-Services/Software/Energy-Optimisation-Software-/WaterTarget/, accessed 7/7/2013.

Printed in Dunstable, United Kingdom